The
Adoptee's Guide
to DNA Testing

The
Adoptee's Guide
to DNA Testing

How to Use
Genetic Genealogy
to Discover Your
Long-Lost Family

Tamar Weinberg

FAMILY
TREE
BOOKS   Cincinnati, OH | familytreemagazine.com/store

Contents

Foreword by Kitty Munson Cooper.. 6

Introduction 8

PART ONE: Your Journey Begins Here

1   Getting Started With Your Search.. 13
    Start your research right with these tips.

2   Proven Search Strategies (That Don't Involve DNA) 21
    Before you turn to DNA, discover what resources such as adoption agencies, social
    media, and birth certificates can tell you about your biological families.

3   The Basics of DNA Testing 33
    Understand the ABCs of DNA with this crash-course guide to the scientific principles
    of genetic genealogy.

PART TWO: DNA Tests and Testing Companies

4   Types of DNA.. 44
    Learn what DNA test is best for you with this wrap-up of the four major types of DNA
    tests: autosomal, X-chromosomal (X-DNA), Y-chromosomal (Y-DNA), and mitochon-
    drial (mtDNA).

5   AncestryDNA... 62
    Join more than seven million users by testing at AncestryDNA, the world's largest
    commercial DNA company. This chapter will show you how to navigate your results,
    from contacting matches to connecting your data to an online family tree.

6   Family Tree DNA 76
    Test multiple kinds of DNA at Family Tree DNA, the only major company to offer
    Y-DNA and mtDNA testing. Learn how to select the best test (and how to interpret
    your results) with this guide.

7   23andMe... 98
    Access your DNA (as well as basic information about your medical history) with
    23andMe. This chapter will show you how.

8   MyHeritage DNA.. 113

See what you can do with a DNA test from the online family tree giant MyHeritage DNA.

PART THREE: **Advanced Tools**

9   Establishing a Biological Connection............................... 124

Reach out to potential relatives! This chapter will teach you how to use mirror trees to figure out how your genetic matches are related to you, plus share some tips for contacting them.

10 Analyzing Your DNA with GEDmatch................................. 135

Take your research to the next level with the analysis tools at GEDmatch.

11 Triangulating Your DNA Data.. 167

Pinpoint relatives on both sides of your family using DNA triangulation, an advanced technique for learning genetic information about a potential relative. This chapter will show you how.

PART FOUR: **Case Studies**

Get motivated! These inspiring stories are real-life examples of how researchers used DNA to solve their family mysteries.

Donna.. 187

Izak... 190

Jack... 192

Kalani.. 195

Kelly.. 201

Marcy.. 206

Sue.. 210

Appendix Frequently Asked Questions................................. 214

Appendix Worksheets.. 224

Index.. 228

FOREWORD

I first encountered DNA and Mendelian theory in high school. Although my career was in technology (where the jobs were), I never stopped following the news about DNA. My husband likes to tell of his college-level biology textbook, which claimed DNA would never be sequenced. How wrong that was!

So here we are today with so many companies that will sample our personal DNA and give us reports. But which test should we use? Which company? When the price came down for personal DNA tests, I tested myself and many family members at 23andMe **<www.23andme.com>** and then Family Tree DNA **<www.familytreedna.com>**. Soon I was inundating my family with e-mails, which got no response.

Family secrets are being unearthed at an astounding rate, but most people have trouble understanding how to use DNA relative match results.

My solution was to start blogging in order to put all the information I was discovering in one place, in the hopes that some of these newfound relatives might access it someday. On the site, I included many how-to posts that I often shared in DNA mailing lists and Facebook groups. And since I am a programmer, I also wrote a few tools to help display and work with the segment match data.

Before long, I had thousands of unique visitors a day and had become an authority on the use of these tests for genealogy, a hobby I have enjoyed for much of my life. When people ask me how I became so knowledgeable on DNA, I tell them that if they were to spend three to four hours a day for five years working with their family's test results, they would become experts, too.

A few years ago, I met one of the founders of DNAAdoption.com **<www.dnaadoption.com>** at a conference where I was speaking. Her explanation of using DNA to find adoptees' birth families fascinated me. Adoptees could use a common ancestral couple or two, then build their trees to find someone in the right place and time. And the shared matches tool at AncestryDNA and other sites provided unprecedented opportunities for adoptees—sometimes enough for those with deep American roots to trace their full ancestry.

The world of adoption searches has changed with the advent of DNA testing. Family secrets are being unearthed at an astounding rate, but most people have trouble understanding how to use DNA relative match results. Now that the database at AncestryDNA **<dna.ancestry.com>** has more than seven million testers, many adoptees open their results there, discover a first cousin or close family DNA match, and barely feel the need to do anything further.

Soon I was taking the occasional adoption case and learning how to use technology to help with unknown parentage cases. And, through my blog, I have gotten to know many wonderful readers and adoptees, including Tamar Weinberg.

Tamar is from an endogamous group, Ashkenazi Jews, and has experience working through knotty DNA issues. She's created this book, a great reference manual to help adoptees in their search and increase their understanding of DNA testing.

This book is an exhaustive compendium of the testing companies and tools available—something that is truly needed for unknown parentage searches, given that no one company or technique works for everyone. What tool is best depends on where your ancestors are from and whether they intermarried extensively (endogamy) or not.

I applaud Tamar for putting this together!

Kitty Munson Cooper
Genetic genealogy blogger
<blog.kittycooper.com>
December 17, 2017

INTRODUCTION

I've been fascinated with the idea of genealogy for as long as I can remember. As a teen during the dawn of the Internet age, I reached out to my family's designated genealogist, my second cousin once removed, Gary, to start a website where the family could meet, learn how they fit in the tree, and stay connected. That never happened, and while I thought about it on more than one occasion, I didn't pursue it. I merely observed Gary's discoveries from the sidelines for years. He started doing genealogy research the same year I was born, so I didn't really think I could catch up.

This book strives to be a resource for those facing tough challenges in their genealogical research: specifically, those who were adopted or (for whatever) reason don't know one or both of their birth parents.

When I was fourteen, Gary self-published a book on our family. I was mesmerized by it, but figured documenting those family relationships couldn't have been too hard. After all, my maiden name—Palgon—is a unique one, and my family already knew we were all related.

I didn't initially appreciate how hard it was for Gary to find all the records, though—they're not so easy to find, and they're all written in foreign languages. Plus, he did this before the digital era. It's amazing what he was able to accomplish before much of the information became available online.

While I was interested in family from a young age, my true passion for genealogy didn't really get ignited until I turned thirty-five. Thanks to social media (especially Facebook), I saw that friends were trying to find family members through the site. One friend in particular, Alan, was often posting to a Jewish genealogy group.

I hesitated to join for quite awhile, considering my Jewish heritage and thinking that Gary had discovered all there was to know about it. After all, my family's story and lineage felt complete, and I never imagined it would be easy to find other lines of my tree. No one knew anything, and some of our ancestors had common surnames. Other names changed entirely, such as my mother's maiden name. (Family story has it that a schoolteacher changed my maternal grandfather's name, and *his* father intentionally changed his name from that of other family members.) With three name changes occurring over a three-generation span, how could I find anyone else?

One day, on a whim, I decided to join the group. Upon my acceptance, I immediately went to work and searched the group's archives for my maiden name—the only one I had considered—and found someone who mentioned this unique and uncommon surname.

This was a "WHOA" moment for me, and later, I learned that one member of the Facebook group, Jason, was my *eighth* cousin. In fact, he was already listed in Gary's book. Since I barely knew my second cousins, the idea that I could trace back to an eighth cousin—and actually put a face to a name—blew me away. Soon, I decided to see what else there was to learn.

Learning about my ancestry using DNA hasn't been easy, however. From DNA testing, religion, and knowledge of my only family's heritage, I know

I'm an Ashkenazi Jew, otherwise defined as one of European Jewish heritage and ethnicity. Ashkenazi Jews like me have a tougher time with DNA research. For one thing, many of us don't have a paper trail that most other Americans can refer to in order to establish their lineage. My family originates from Poland, Russia, Hungary, and Lithuania. While I'm lucky enough to have gotten records from some of those countries, it will be difficult to find anything from the others. During the Holocaust, many records of the Jewish people in Europe were destroyed—even cemeteries, where stones were pulled out of the ground to build roads (or just to be desecrated).

Another challenge is simply the language barrier. Unlike Gary, I don't feel comfortable communicating with others in Polish—Gary has that method down pat. Thankfully, we're in the Google Translate era. (I don't know if Google is butchering my English when I communicate with foreign entities, but I'm still getting understandable responses back from people, which is progress!) However, response times can take days or even months. Even with that, the language disconnect is discouraging, especially when you don't get the exact answer you're seeking and they request more information. Then, it's back to Google Translate. All I have to say is, "Thank you, Gary."

Perhaps the most difficult part is the endogamous nature of our people. Endogamy is the phrase that refers to the intermarrying within the tribe. Put simply, most Ashkenazi Jews are related to each other, and some say we're all fifth, sixth, seventh, or eighth cousins. If I compare my DNA data to the databases consisting of European Jews, I'll find about seven thousand matches. And if I contact one of them to establish a common ancestor, I won't get much new or useful information. They know their ancestry and I know mine, but we don't go far back enough to establish a clear path.

Think about it. We have:

- two parents
- four grandparents
- eight great-grandparents
- sixteen great-great-grandparents (second)
- thirty-two great-great-great-grandparents (third)
- sixty-four great-great-great-great-grandparents (fourth)
- 128 great-great-great-great-great-grandparents (fifth)

And so on, until we get to 16,384 great-great-great-great-great-great-great-great-great-great-great-great-grandparents.

Let that sink in for a minute. More than sixteen thousand people, twelve generations ago, could have helped make you into the person you became. But in the European Jewish world, it's likely significantly less, because cousins married cousins (making a mess for later genealogists to sort out). They call the repetition of a single person in a family tree pedigree collapse—when relatives marry each other, and the branches of your family tree become tangled.

Because of this, I know the road to finding your family isn't always straight. Fortunately, DNA helped me sort out some of my complicated family history, and my challenges motivate me to help others learn about their own histories. Going on a DNA path can feel overwhelming, so we're here to discover the science and how to make sense of your discoveries.

This book strives to be a resource for those facing tough challenges in their genealogical research: specifically, those who were adopted or (for whatever) reason don't know one or both of their birth parents. We'll discuss strategies to help you reconnect with your birth family using DNA, plus what DNA tests are available to you and how you can use and interpret your results. I'll also share stories (both good and bad) of real people discovering their biological families to give you hope and remind you that you're not alone.

Perhaps you're not adopted. Perhaps you know no one in your family who was given up for adoption or who was adopted. Perhaps you just want to cast a wider net within the family you already have and know. This book serves to meet the intellectual curiosity of those who would be interested in DNA and general record searching and what you could find as a result. I'm hoping that one day the information you learn in these pages will teach you how to find your families, too, and to give you the motivation to move forward.

Finally, for more on DNA tests and this book, please visit my website.

Tamar Weinberg
Writer and genealogy enthusiast
<tamar.com>
December 17, 2017

PART ONE

Your Journey Begins Here

1

Getting Started with Your Search

Perhaps you've known all your life that you were adopted. Perhaps you recently discovered that you were adopted. Perhaps you just feel that you're out of place within your family dynamic and you remain unconvinced that your birth parents are truly your birth parents. Perhaps you'll take a DNA test and find out that you aren't who you always thought you were.

You may have asked many questions along the way. After all, we all want to know where we came from. Do I have brothers or half-brothers? Do I have sisters or half-sisters? Aunts? Uncles? What were they like? What were my grandparents like? What is my real ethnicity? Are there medical issues of which I should be aware? It is human nature to want understand where you come from and what your family situation is. And asking "Who am I, and where do I come from?" is perfectly natural and reasonable.

Most people seek birth families—despite all of these questions—to help fill a void. Many need a sense of closure to address the gaping hole in their hearts. They often seek relationships with new family members, while not throwing away the relationships that preceded them. Doing so doesn't change the past or unravel the threads that have nurtured them—nor, likely, do they want it to. Their intentions are almost always pure, and most don't want to disrupt a birth family that has since moved on.

The desire to learn more about birth families can affect anyone touched by adoption, from an adopted child seeking his genetic ancestors to the parent who had to give up her child and wants to reconnect. Perhaps a sibling wants to find a child given up under less-than-ideal circumstances, an individual knows he has a long-lost half-sibling, or a man suspects a one-night stand resulted in a baby. Or maybe you've discovered the man who raised you isn't your birth father, which is called a non-paternity event (NPE).

Regardless of who or why, someone might be curious and wants to know what became of the human being who was brought into the world and put into another family's arms. Or, from the adoptee's perspective, wants to know who his parents and other family members are. Some just want an answer to the question, "Why?"

To find these important answers, many adoptees and families involved in adoption have come to rely upon various tools and tactics, and technology has evolved into a multitude of resources. In less than a decade, DNA tests have become relatively affordable for the general public. These new tests have amazing, highly accurate features to help you find birth relatives, including databases of matches, ethnicity information, and (on some tests) statistics about how your genetics compare with those who share DNA with you. Better yet, these tests are widely available and boast databases of millions of people, many of whom are also looking for clues about their own families.

At time of publication, four major players (AncestryDNA **<dna.ancestry.com>**, 23andMe **<www.23andme.com>**, Family Tree DNA **<www.familytreedna.com>**, and MyHeritage DNA **<www.myheritage.com/dna>**) have cornered this market, but more and more companies are following suit and offering similar services that enable test-takers to establish fairly accurate relationships. In fact, DNA science will likely, in time, advance so far that you're able to take and process a test from the comfort of your own home.

Before the advent of genealogy companies, researchers had to resort to finding information through an adoption agency or birth certificate, two imperfect sources that we'll discuss in more detail in chapter 2. Name changes or fabricated/inaccurate information can corrupt these resources, providing false information. With these traditional methods, one's success was a matter of searching—perhaps with the help of a licensed investigator, a phone book, and certified mail. However, a person could be unlisted in the phone book, or he moved and failed to update his contact information.

Modern technology such as social media has made these conventional methods somewhat more accessible, but even these have significant drawbacks. The person you're looking for may not have an online presence, or perhaps he rarely updates it and won't receive a message. And even if you do find relevant information, it may not provide the answers you need. The possibilities go on and on.

This is exactly why so many who have been involved in the adoption process have turned to publicly available resources such as DNA testing. No matter what you know about your birth family's history, DNA doesn't lie. If you can document a relationship through DNA (and you draw the correct conclusions from your results), you can usually trust it more than any paper trail you have.

However, the information at your fingertips can be overwhelming. Not every DNA test is the same, and each DNA website has its own pros and cons. How do I decipher my results, and how can I make sense of the data to establish a relationship? What happens if the connections aren't so strong? How accurate are these services? What should I expect? Is my data safe and secure in the hands of a third party?

Those questions, my friend, are why we are here. This book will walk you through how to sift through these questions to find your birth family using DNA. But first, let's discuss what you can realistically expect to discover in your research.

Why Should I Test my DNA?

By taking a DNA test, you are enrolling in a database of other test-takers who also want to know more about who they are and where they come from. DNA research is only as strong as the data comprising it, and who takes the test is an important factor in how useful your results will be.

The more people related to your research who test, the better your chances are for finding someone who you may be related to. If you suspect you are adopted or a family member is adopted, test. If you have a genuine curiosity about where you fit in the context of relationships with others, test. And—most of all—if you want to learn your genetic makeup and find family members, test.

Testing more people helps the genetic genealogy community at large as well. By building up a database of test-takers, we can help those who pursue genetic testing for other reasons. Some are interested in health information, for example, and test on sites that provide that context. They may even test on another site and import their data to a third-party website (such as GEDmatch **<www.gedmatch.com>**) to gather whatever health information is available. We'll go into this implementation of DNA data in chapter 3.

Why else do people test their DNA? Many are swayed by the ads they see online and on TV that emphasize the tests' ability to provide ethnicity estimates, or a breakdown of which part(s) of the world your family comes from. While these ads are great at attracting customers to these testing companies, they oversimplify the test's features, and many people don't know just how much other data can be gleaned from a DNA test. Those who do DNA testing for these reasons are generally less interested in reconnecting with DNA relatives, making it difficult for you to establish contact with them.

Another group of DNA testers is interested in genealogy and wants to reconnect with family in some form or another. After all, genetic genealogy is a tremendously useful resource for building a family tree. Some groups, in particular, have more difficulty using DNA than others; we'll go into some of these challenges throughout this book when we touch upon the topic of endogamy, or generations of intermarriage within a community.

Regardless of your intention to test, be willing and able to respond to people who have reached out, whether or not you know the answer to their questions. Those heavily

LEVELS OF COUSINHOOD

First cousin twice removed? Second cousin once removed? What's the difference? If you're just getting started with learning about your family history, all the terminology can be confusing. But in reality, the concept is simpler than you might think—and understanding levels of cousinhood is critical for identifying genetic relatives.

To put it simply: The degree of cousin you are (first, second, third, etc.) refers to the most recent ancestor you and your relative share, while the "level of removedness" (once removed, twice removed, etc.) describes how many generations are between you and the other person.

Let's start with the basics. Your first cousin, as you probably know, is a child of your parent's sibling—your aunt or uncle. This first cousin shares a common ancestor with you, and that ancestor is a grandparent to both of you.

And how are you related to your first cousin's son? You still share a common ancestor (your and your first cousin's grandparent), but that individual is the child's great-grandparent. You and your cousin's son are *removed* a generation, making you first cousins once removed.

So what's a second cousin? These are two individuals who share a set of great-grandparents, but different grandparents. Usually, this means your second cousin is the grandchild of your grandparent's sibling. You can tack on the "removed" designation as necessary. For example, your second cousin's child would be your second cousin once removed.

This chart will help you calculate how you're related to various individuals in the family tree you're building. Start at the box that says Self, then move up the chart to find the ancestor that you and this relative share in common. Then, count down from that ancestor, moving down one box for each generation between this ancestor and the relative you're trying to match. The box you land on will be the relationship between you and this relative.

This chart also shows how much DNA you can expect to share with each kind of relative, both in terms of percentage and number of shared "centimorgans." We'll discuss all that DNA data in later chapters—for now, just focus on the family relationships.

Let's look at an example. Let's say a woman named Daphne is trying to figure out how she and her first cousin's child, Edgar, are related. Daphne and Edgar both share a man named Albert as an ancestor; Albert is Daphne's grandfather and Edgar's great-grandfather. So Daphne will use Albert as a reference point. She starts at Self and counts up to grandfather. Since Albert is Edgar's great-grandfather (three generations before Edgar), Daphne would then count three down from Albert's box. The result is first cousin once removed. The chart works in reverse as well—start with Edgar as Self and move up to Great-grandfather Albert, then down two generations (since Daphne is Albert's granddaughter). The result is the same.

Learn more about the intricacies of calculating cousinhood at FamilyTreeMagazine.com **<www.familytreemagazine.com/premium/how-to-calculate-cousinhood>**.

How to Calculate Cousinhood

Follow these steps to figure out what kind of cousins you are with a relative:

1. Identify the most recent ancestor you share with your relative, and how that ancestor is related to both you and to your relative.

2. Find the ancestor on the chart (such as your parent, grandparent, great-grandparent, etc.).

3. Count down one box for each generation between that ancestor and your relative. The box you land on specifies your relationship with the relative, and how much DNA you share with him or her.

The shared DNA data can help you estimate your relationship to a genetic match. Note this chart doesn't show double cousins or half siblings, both relationships with about 25 percent shared DNA (roughly 1,700 cM).

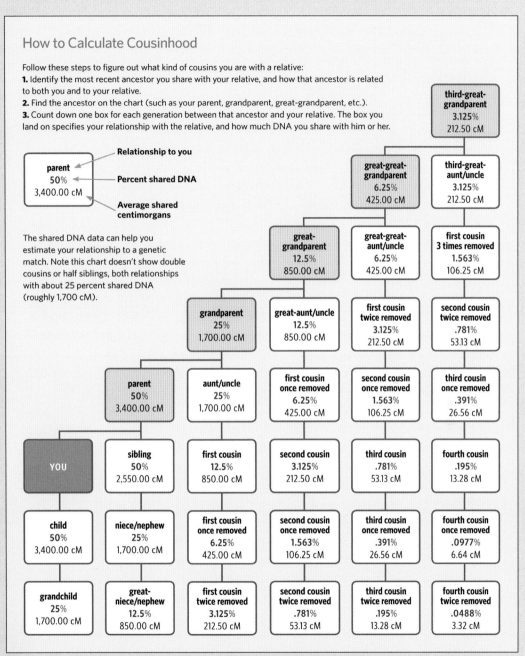

Use this chart to calculate how you and another person (say, a genetic match) are related.

invested in genealogy, myself included, will do whatever they can to help adoptees reconnect with their families. That collaboration is what makes the genetic genealogy community so helpful.

Aligning Expectations

Your results may vary when it comes to DNA testing. A DNA test could almost instantly connect you with an immediate family member, or it could take years before you break the ice and make traction with your search. DNA is a fantastic way to establish a familial relationship, but it is not necessarily a quick process.

For example, a New York woman named Melissa (name changed) wanted to find information about her birth family, who likely was from the American South. Melissa was skeptical of DNA's utility, but she tested anyway on the off-chance that it could provide useful information about her parents. Years later, she had all but given up after not making any research finds—that is, until one day she found a full-sibling match. This woman, another New Yorker, was born eleven months after Melissa and was also adopted. Doctors hypothesize the two are twins given incorrect birth dates. Melissa's original quest (to find her birth parents who she believed to live hundreds of miles away) led to a sister living fifteen minutes away from her in New York. You never know where DNA will take you.

Furthermore, your DNA results may surprise you. For example, my father-in-law (who was not adopted) received a second-cousin match that he swore couldn't be correct; after all, he'd never heard of her before. But after the match confirmed all the parents' and grandparents' names, we found that this woman was, in fact, his second cousin twice removed. We were fortunate that both sides of this equation were willing to share information in order to establish a direct link between unknown cousins. As a result, you should keep an open mind, and consider the ways in which DNA tests can be correct.

Perhaps most important, regardless of whether you test via DNA or not, be open to the possibility that the end result of your research might be unexpected. You may get lucky and match with a like-minded immediate family member who wants to rekindle that lost relationship. But the individuals you're looking for may not yet be in a genetic database, or they might not want to revisit that part of their history. More distant relatives, unaware of the situation, might not want to engage with you or admit that an adoption even occurred in the family. Or perhaps your research uncovers a non-paternity event (e.g., a man turns out to not be the genetic father the family thought he was) that upsets family dynamics.

Needless to say, DNA can unearth unexpected discoveries, and you need to accept this before you begin your journey. You'll never be able to close Pandora's box. In one case I worked on, we learned that the subject's assumed full siblings were actually half

CONFRONTING SKEPTICS

Despite the power and potential of DNA (the science, when correctly applied, is sophisticated enough to confirm relationships), many people are still skeptical about genetic genealogy. A family member may refuse to accept discoveries made through DNA research, especially in adoptions where family members reject the same kind of data provided by shock reality shows on TV.

To appease those skeptics, you could have a local lab confirm those results, though (depending on the quality of local genetic services) these labs could produce false positives or negatives. Also, local labs don't give you access to the wider population like the testing companies we discuss in this book, and you'll likely need that to find family members on your own. Or you could provide your skeptical family members with more detailed information on the testing companies and their methodologies.

siblings—her brother and sister shared the same father, but she had a different father. In other cases, people have discovered half siblings they've never even heard about. It can be frightening to assume you know all about your family line, but the DNA proves otherwise. Some people may need time to process and accept these new genetic relatives (if they're receptive to them at all). As a result, take care when approaching newly discovered genetic relatives. If you find a match, make sense of it before you reach out—throughout this book, you'll learn how.

Discoveries may surprise you and may even disappoint you. You might find an ethnicity you don't especially like, or that someone you hate is actually a distant cousin. Have an open mind and embrace who you are, as your heritage is not something you can control—and your diverse background is actually a beautiful thing!

One viral YouTube video brings the importance of DNA testing to life. Momondo, a global travel search website, created a study in which sixty-seven people from around the world discovered where their genetic heritage comes from **<www.youtube.com/watch?v=tyaEQEmt5Is>**. Each was interviewed before learning their results, sharing their affinities for and aversions to different nationalities. When the results came in, not only did they discover unexpected ethnic origins, they also discovered that some of them had family in the room. The video truly encapsulates what you can discover through the pursuit of your heritage.

Before you test, set aside your preconceptions and prepare to embrace your true self and be proud of who you really are. I know I am proud of all the family I've come across in my research, no matter how different they are from me and my ancestors. I can confidently say that my family, like every family, like just about every other, has undergone significant change from generation to generation and from sibling to sibling—religiously, spiritually, and philosophically. I embrace the diversity that defines my family today—though, in all honesty, not everyone in my family does. Still, I wholeheartedly suggest that you prepare to do the same. The results, in many cases, are eye-opening and paint the picture of who you truly are. Let's get started!

2

Proven Search Strategies
(That Don't Involve DNA)

s we mentioned in chapter 1, adoptees and others with unknown parentage often try to search for their family members through more traditional channels before turning to DNA. After all, DNA technology is relatively new, and adoption cases of decades past had to be settled using other means (if they could be settled at all). DNA can help you identify and possibly connect with birth family members, but (if possible in your state) you should potentially look at other pieces of data that might get you there even faster.

So where can you start before consulting DNA? Normally, we begin with what you already know about your birth parents, followed by search engines and social media. These early stages of research could be as easy as finding a birth family member's social media account, but other instances may not be so straightforward.

In this chapter, we'll discuss how to find genealogy records and take advantage of other search strategies before you dive into DNA testing.

RESEARCH TIP

Don't Start with DNA

While it can be tempting to jump into DNA testing right away, start with traditional research techniques and records before turning to genetic testing. DNA testing's usefulness depends on who has tested so far, and you could wait a long time before connecting with a DNA match in the database.

Birth Certificates

Birth records are amongst the most important kinds of records in your search, as they can easily connect you with your birth family and provide the missing link to your family history. Your ability to find these records will vary based on where you're searching, as each state has laws regarding access to birth certificates. Some states have loose access policies, while others are more restrictive.

Your birth certificate (image **A**) will often provide some important pieces of information: your name, your mother's name, and your location and date of birth. Birth certificates for adoptees may not include the father's name, which may be a disappointment for those searching exclusively for their birth fathers. Even so, the information from a birth certificate could be enough to give your research some general direction. Sometimes, information from these records just isn't factual enough to work off of, as we'll learn when we read Kalani's story at the end of this book.

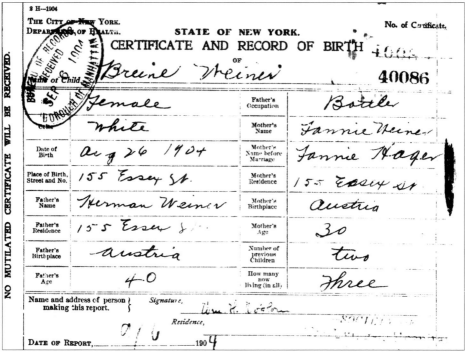

Image A. If you can find and access them, birth certificates (like this one from 1904) are critical resources for finding birth families. They can provide you with the names and birthplaces of both biological parents, plus other interesting details such as occupation. Unfortunately, many states have laws preventing individuals from accessing these records.

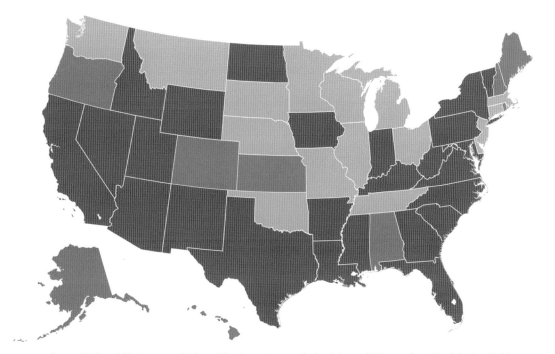

Image B. The ability to access birth certificates varies greatly by state, and this map from the Adoptee Rights Law Center, categorizes these laws. Red and yellow indicate states with restricted access to birth certificates in some or all instances, and green indicates states in which individuals can freely access birth certificates. Refer to the Law Center's website for the latest version of this map.

Keep in mind that birth certificates may or may not have the information you need. Not all birth certificates are accurate, adding another potential wrinkle to your research. While this happens only in rare instances, consider the possibility that even birth certificates might not hold all the correct answers you need.

If you were born in a state in which a court petition is required to access your birth certificate, you'll receive either an unredacted birth certificate or an amended birth certificate. The former gives you all the information you need, and some states offer this without restriction. Unfortunately, more states are limited in the information they freely allow adoptees to obtain.

All fifty US states have different laws concerning access to birth certificates. Some simply require you to file a formal request, while others keep these records under lock and key. The map in image **B** shows three categories of adoption laws as of this book's publication. Red indicates states in which birth certificates are restricted, while yellow represents states that allow individuals to access their records in specific circumstances. Green indicates places in which adoptees can freely access their birth certificates.

A WARNING ABOUT ESTABLISHING CONTACT

With the tips in this chapter, you're hopefully getting closer to finding one of your birth relatives. But before you initiate contact with any of them, I'd advise you to pause and consider your strategy—as well as the other person's feelings. Most people won't be prepared to hear from a long-lost child or parent, and some won't want to engage with you. You'll need to be tactful in how you reach out to these new family members.

First, find a delicate way of delivering information about your relationship to this person. Start more generally, then (as you grow more comfortable with each other) reveal the more specific—and, potentially, more emotional—details. A good lead-in to the conversation might be your genealogy research storyline. You can say you're simply hoping to learn more about your genealogy, and perhaps share some of your other findings. For example: "Hi, my name is [name], and I need your help. I'm working on a genealogy project, and I think we may be connected. You can reach me at [phone number] or [e-mail address]." This is the approach I took with a cousin who had a son he didn't know about. I began with "Hi, I'm your cousin," and I name-dropped a few people from my research. A mutual relative who this new family member already knows might also help you break the ice.

Once you have this person's attention, tell him what you know. Perhaps you can provide insights into your birthdate and birthplace—that, alone, may suffice to give them the information they need to want to get to know you. Returning to the previous example: Once my subject and I had started a dialogue, I explained that I had big news to share. I gave him the name of the child's mother. He said he knew her, and that's when I told him that I think he fathered her son. To my surprise, he confirmed, and father and son later reunited and began learning about each other.

Don't be too forceful and don't press the matter. But do tell them that you want to get to know them. "I believe you are my father/mother" can open the floodgates of emotions for both of you, which could lead to a successful reunion and newer relationships that emerge from there.

Finally, remember that contact is a two-way street. You need to be able to contact the individual, and he needs to be able to contact you. So when you make contact, make sure that he has a way to get back to you. In chapter 9, we'll talk more about this and feature some sample scripts you can use when connecting to your birth family.

Find the latest information regarding adoption and records access laws by state or location at the Adoptee Rights Law Center **<adopteerightslaw.com>**. The American Adoption Congress also has a chart describing laws by state **<www.americanadoptioncongress. org/state.php>**.

As you might expect, accessing birth certificates from different countries also varies. Check the website for your country's government and/or archives to learn more about its records access.

Adoption Registries

Your state will likely also have an adoption registry where you should enlist yourself. Adoption registries help facilitate possible reunions between birth families, particularly useful in states that have closed adoptions (adoptions in which parental information is withheld). By using these registries, you may be able to obtain additional information about your birth family. They also can provide identifying information where available, non-identifying information, or medical/psychological information. With luck, you may even be able to reunite with family members already looking for you through an adoption registry!

You have several types of adoption registries to choose from, including government-run, non-profit, and private business entities. In most cases, you'll just need to provide your birth date, birthplace, gender, and name, though be advised that some may charge a fee.

Note that different adoption registries have different policies about disclosing information. State laws prohibit some governments from sharing information, and non-government-run adoption registries may only be able to provide information based on mutual consent. In other words, the adopted child and the birth parent both have to agree to allow information to be shared. Without permission, the adoption registry may have little to provide to the person performing the search.

Age is another limiting factor in your success with adoption registries. In general, you have to be at least eighteen years old to request information from a registry. If you (or an

---

RESEARCH TIP

Go Back to the Beginning

You may also consider finding the agency that orchestrated the adoption. By reaching out to this organization, you may be able to receive non-identifying information about your birth parents or insight into where to look next. Your actual birth records may be sealed and inaccessible, but a birth index might be able to provide a few facts about your birth parents.

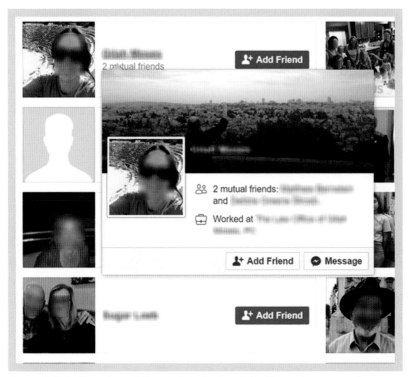

Image C. Mutual friends can help make introductions to family members—assuming you can find them using tools such as Facebook.

adoptee you're trying to help) are under eighteen and require more immediate answers, consider skipping to the DNA route. Some sites allow minors to test with the permission of a legal guardian; check the site's terms and conditions.

In addition to those run by states, more general adoption registries can be of use to researchers. Check these out:

- Adopted.com <www.adopteeconnect.com/ReunionRegistry.org>
- Adoption Reunion Registry <registry.adoption.com>
- FindMe.org <www.findme.org/index.cfm?fuseaction=Main>
- G's Adoption Registry <www.gsadoptionregistry.com>
- International Soundex Reunion Registry <www.isrr.org>
- QuickBase Adoption Database <adoptiondatabase.quickbase.com>
- Reunion Registry <www.reunionregistry.org>

Wherever possible, try to access your records by reading more about state law, consulting registries in specific states, or accessing record request forms directly. Check out the Adoption Records List at the end of the chapter to find links for individual states.

Social Media

In addition, a number of Facebook groups serve adoptees trying to reconnect with birth families, and these can provide resources to aid your search (and, just maybe, lead you to family members). DNA Detectives **<www.facebook.com/groups/DNADetectives>** and Search Squad **<www.facebook.com/groups/searchelpers>** are two great general resources. These groups comprise volunteers who can either guide you toward records access or help you interpret your DNA results. The DNA Detectives group, in particular, is more collaborative, featuring multiple contributors who can answer questions and/or provide supporting evidence. Likewise, the Search Squad assigns a volunteer called a "Search Angel" to help crack your case, hopefully working to reunite your family.

A number of other, smaller groups can also help you find your loved ones on a regional or statewide level. Simply search Google or the various social media platforms to find and join one. Other groups can help you translate documents or access obituaries, two other tasks important to your search. Don't be afraid to ask questions; the members of these groups are volunteers who genuinely want to help or who have been there, done that, and know what you're going through.

Think broadly about the kinds of social media platforms your potential family members might use, as you never know where you might discover one of your genetic relatives. Be sure to search for your person of interest on Facebook **<www.facebook.com>**, Twitter **<www.twitter.com>**, Instagram **<www.instagram.com>**, Google+ **<plus.google.com>**, LinkedIn **<www.linkedin.com>**, and Pinterest **<www.pinterest.com>**. Your relative might also have a blog, where you can learn about their life or potentially find contact information.

Reaching Out to Relatives on Social Media

After consulting birth certificates and adoption registries, perhaps you found a potential relative—or, at the minimum, at least a name. The next step is to research their digital footprint. Facebook is an obvious first step, but connecting with a long-lost relative isn't as simple as finding and messaging them. Messages from an unknown person could be ignored or go unseen for weeks, months, or even years. How can we more practically approach potential relatives?

Let's take a step back. Fortunately, you're more connected with everyone on Facebook than you think. In early 2016, Facebook found that (on average) users have just 3.5 degrees of separation between them. To put it another way: Most Facebook users are connected via a friend-of-a-friend-of-a-friend or closer. While you may not necessarily know the person you want to speak with, you may have a friend in common with her—or maybe a friend who has a friend who has a friend in common with her.

You can test this theory by exploring mutual friends. If another user (for example, a potential relative) has made his friends list publicly available, you can hover over each of his friends' names to see if a mutual friend (or friends) shows up. You can see this in image **C**, in which my mouse is hovered over the name of the last person on the mutual friends list. As it turns out, I have two mutual friends with her.

If you have a mutual friend with the desired contact person, ask her to make an introduction or even get an e-mail address or phone number. Then, reach out. The best way to do that is under the guise of researching your family genealogy. Never start a conversation with a blood relative by disclosing your believed relationship—that information may be too shocking for your newfound relative, and it's usually best to wait until you've built up an acquaintanceship before disclosing those details. We'll discuss how to reach out to family members in more detail in chapter 9.

Unfortunately, you can't always use the mutual friends tactic. Some Facebook users do not make their Friends list publicly accessible. If this is the case, you won't be able to go through their Friends list, but you can scroll through their page and photos/shares posted. Facebook doesn't hide the number of likes and comments on a post; hover over any of those names and see if there's a common name or a name with friends in common. This can be an arduous (but worthwhile) task.

Keep in mind: Even if all of these strategies have failed, you've at least tried. And you still have many more avenues to explore!

Online Search Tools

Like you, I sometimes cannot get in touch with the people I want. While Facebook is the likely place to check, there are other ways to get closer to finding people, assuming they have some semblance of a digital footprint.

A number of search engines can find whatever digital footprint exists for others. Most notably, Pipl **<www.pipl.com>** and Spokeo **<www.spokeo.com>** can give you a general idea of a person's location, background, habits, and more. Spokeo will give you additional information for a fee, and Pipl curates the person's social profiles in one centralized location. You may also want to try BeenVerified **<www.beenverified.com>** or Intelius **<www.intelius.com>**. Both are subscription services that often provide a great deal of context regarding your search target. These sites may be a bit costly, so don't use them unless you're confident you're on the right track or you have a whole list of names to research. Members of the DNA Detectives and Search Squad Facebook groups can also assist you here.

I have also used a number of people-based tools designed for business contexts that provide you with a person's e-mail address, if available. While you may not have this information readily available now, you will find it useful later when e-mail addresses are made available through DNA databases.

Discoverly **<discover.ly>** is one of these tools. This service searches various platforms for a name and provides social media information about a person with that name. Discoverly gives Facebook information and data extrapolated from various sites, but note that it is not always correct. For example, if you are looking up an Adam Smith in Kentucky, Discoverly may provide the wrong person, Adam Smith from Oregon. Discoverly works only on the Google Chrome browser, and (though free to access basic data) you'll need to pay a small subscription fee to view more results.

If you find someone on a social network like Twitter or Instagram, you can also search his username in Google; perhaps he uses it elsewhere. Be sure to put his name, including or excluding maiden name or middle name/initials, in quotes. Sometimes this won't turn up anything, but it's worth it to try. If something does show up, there may be an e-mail address there too or a contact button. Use it!

ADOPTION REGISTRIES LIST

Note that these website URLs are subject to change. If you find a broken link, browse your state government's website to find information about accessing the documents you need in your research.

Alabama: <www.alabamapublichealth.gov/vitalrecords>

Alaska: <dhss.alaska.gov/dph/VitalStats/Pages/default.aspx

Arizona: <www.azdhs.gov/licensing/vital-records/index.php#adoption-home>

Arkansas: <www.healthy.arkansas.gov/programs-services/program/certificates-and-records>

California: <www.cdss.ca.gov/Adoptee-Information>

Colorado: <www.colorado.gov/pacific/cdphe/adoption>

Connecticut: <www.ctfosteradopt.com/fosteradopt/cwp/view.asp?a=3795&Q=564922>

Delaware: <www.dhss.delaware.gov/dph/ss/files/adopted.pdf>

Florida: <adoptflorida.com/Reunion-Registry.htm>

Georgia: <www.ga-adoptionreunion.com>

Hawaii: <www.courts.state.hi.us/docs/1FP/1FP770.pdf> and <www.courts.state.hi.us/docs/1FP/1FP767.pdf>

Idaho: <healthandwelfare.idaho.gov/tabid/1504/Default.aspx>

Illinois: <www.dph.illinois.gov/topics-services/birth-death-other-records/adoption>

Indiana: <www.in.gov/isdh/20371.htm>

Iowa: <dhs.iowa.gov/adoption-records>

Kansas: <www.dcf.ks.gov/services/PPS/Pages/Adoption-Records-and-Search.aspx>

Kentucky: <www.kyadoptions.com/adoptees.html>

Louisiana: <www.dcfs.louisiana.gov/index.cfm?md=pagebuilder&tmp=home&pid=116>

Maine: <www.maine.gov/dhhs/ocfs/cw/adoption/reunionregistry.htm>

Maryland: <dhr.maryland.gov/adoption/search-contract-and-reunion>

Massachusetts: <www.mass.gov/eohhs/docs/dph/vital-records/pre-adoption-app-form.pdf>

Michigan: <www.michigan.gov/mdhhs>

Minnesota: <www.health.state.mn.us/divs/chs/osr/adoption.html#adoptee>

Mississippi: <thenationalcenterforadoption.org/wp-content/uploads/2014/02/mississippi.pdf>

Missouri: <health.mo.gov/data/vitalrecords/adopteerightsact.php>

Montana: <dphhs.mt.gov/vitalrecords/contacts.aspx>

Nebraska: <dhhs.ne.gov/children_family_services/Pages/adoption_searches.aspx>

Nevada: <dcfs.nv.gov/Programs/CWS/Adoption/Guide/NVAdoptionReunion>

New Hampshire: <sos.nh.gov/Pre-Adoption.aspx>

New Jersey: <www.nj.gov/health/vital/adoption/vital-record-law-changes-faqs>

New Mexico: <nmadoptionsearch.com/procedure>

New York: <www.health.ny.gov/vital_records/adoption.htm>

North Carolina: <www.ncdhhs.gov/divisions/dss>

North Dakota: <www.nd.gov/dhs/services/childfamily/adoption/disclosure.html>

Ohio: <www.odh.ohio.gov/vitalstatistics/legalinfo/adoption.aspx>

Oklahoma: <www.okdhs.org/services/postadopt/Pages/default.aspx>

Oregon: <www.oregon.gov/dhs/children/adoption/Pages/registry.aspx>

Pennsylvania: <www.adoptpakids.org/Documents/PAIRBrochure.pdf> <www.health.pa.gov/MyRecords/Certificates/Pages/Adoption.aspx>

Rhode Island: <www.health.ri.gov/records/for/adultadoptees>

South Carolina: <www.southcarolinaadoptions.com>

South Dakota: <dss.sd.gov/childprotection/adoption/registry.aspx>

Tennessee: <www.tn.gov/dcs/program-areas/fca/adoption-records.html>

Texas: <www.dshs.texas.gov/vs/reqproc/adoptionregistry.shtm> <www.dfps.state.tx.us/Adoption_and_Foster_Care/Adoption_Registry>

Utah: <vitalrecords.utah.gov/adoption>

Vermont: <dcf.vermont.gov/vt-adoption-registry>

Virginia: <www.dss.virginia.gov/files/division/dfs/ap/intro_page/guidance_procedures/records.pdf>

Washington: <www.doh.wa.gov/LicensesPermitsandCertificates/BirthDeathMarriageand-Divorce/Adoptions/AdoptionLawSHB1525>

West Virginia: <www.wvdhhr.org/bcf/policy/adoption/Adoption_Policy.pdf#page=133> (sections 13.1 to 13.4)

Wisconsin: <dcf.wisconsin.gov/adoption/search>

Wyoming: <docs.google.com/file/d/0B6DSpyyE-UESUXV1SXITODhQZkk/edit?pli=1>

Washington, DC: <www.dcd.uscourts.gov/adoption-petitions>

3

The Basics of DNA Testing

f conventional search strategies have proven fruitless (or you want to dive deeper), you might next turn to DNA testing. Fortunately, DNA's accuracy lends itself to being an extremely powerful testing method, and testing with your DNA has both short- and long-term benefits for your research. If you're in a testing company's database, you'll make it much easier for other people to find you.

Because your unique DNA defines who you are, DNA testing can be an invaluable tool for adoptees and others with unknown parentage. For a relatively low cost (usually between sixty and one hundred dollars), you can submit a DNA sample to a testing company, which then logs your DNA data into a system and matches you with other test-takers who share segments of your DNA. In addition to providing interesting information about your geographical genetic breakdown, these tests can provide you with unprecedented amounts of information about potential blood relatives. And since millions of people have tested already, you will almost certainly find matches, albeit some who aren't that closely connected to you. Each and every one of these connections is valuable for you in your quest.

Before we tear into DNA testing strategy, we need to understand the science behind genetic genealogy. This is key, as I have seen people both underestimate or overestimate the strength and accuracy of DNA-based relationships, and these misconceptions have serious consequences in how you identify and even interact with potential relatives. By understanding the science and terminology, you can also better reach out to experts and other test-takers to deepen your research when you've hit a brick wall.

This chapter will give you a crash-course in genetic genealogy, from the basics of "What is DNA?" to how to submit a DNA sample.

WHEN DNA FAILS

In some communities, autosomal DNA results are less accurate due to cultural practices, such as a tendency for people to marry only within their community due to cultural, societal, and/or religious traditions. This practice, known as endogamy, results in all people who come from these populations being interrelated, often in multiple ways. Endogamous populations include Jews (primarily Ashkenazi), Polynesians, Amish, Mennonites, French Canadians, Acadians (Cajuns found in current-day Nova Scotia), Newfoundlanders, Arabs, and residents of islands who were confined to their boundaries for generations.

As a result, DNA testing companies can overestimate the number of matches for test-takers who come from endogamous communities. This can cause a discrepancy between DNA results and your family tree, muddying the true familial connections between test-takers. For example, someone whose DNA results estimate to be a third cousin is more than likely a fifth through eighth cousin. Due to intermarrying in the community, two individuals are related both on the mother's side and the father's side.

I personally am part of an endogamous community: Ashkenazi Jews. Whereas many people have five hundred fourth-cousin matches or less, I have nearly five thousand—and my grandmother has eleven thousand. Are they all fourth-cousin matches and worth researching? Certainly not. One of my strongest third-cousin matches has no documented close connection to me whatsoever and looks equally connected to me through my mother and my father. I likely will never find out how she is connected to me simply because it is too hard to establish the relationship using records, and the DNA convolutes and clearly overestimates the relationship. Likewise, my "fourth-cousin" matches are more likely seventh or eighth cousins, and it will be nearly impossible to establish a true connection to them due to the absence of records and name changes. Keep this tendency in mind if you know your ancestors' communities were endogamous.

In short, in endogamous cultures, "Everyone is related," and this can skew your DNA results. In fact, a report from Columbia University found that Ashkenazi Jews can trace their ancestry back to a group of just 350 individuals, making all of us at least thirtieth cousins **<www.nature.com/ articles/ncomms5835>**.

What is DNA?

Wait! Don't skip this section. The basics of DNA is not as scary (or as dull) as you think. You don't have to be a DNA expert to understand the basics or to understand and use your results. DNA is just one part of biology, and the subject is even fun once you study it for awhile! DNA (as it applies to our particular research) is relatively easy to understand, and so are its analysis tools and real-world applications. Give thanks to modern technology for making sense of all this biology!

Unless you have an identical twin, no one in the world shares 99.9 percent DNA with you. Your DNA comes from your ancestors who preceded you, and genetic material from your ancestors lives on in siblings and cousins who shared DNA with them. The more distant the relationship, the less DNA you share with these cousins. These test results can tell you just how strong the relationships may be.

First things first: Cells are the basic structures of a living organism. A human body contains billions or even trillions of cells, each composed of several different parts. For the purposes of genetic genealogy, the most important components of a cell are mito-chondria, the nucleus, DNA, chromosomes, genes, and alleles. Here's a quick definition of each, in turn:

- Mitochondria are the "powerhouses" of the cell that produce energy. They contain unique DNA that's passed along from mother to child, and certain DNA tests ana-lyze just this genetic material. We'll explain how mitochondrial DNA can help us with our research in chapter 4.

- The nucleus is the command center of the cell, and is involved in carrying out instructions for cell growth, reproduction, and death.

- DNA, otherwise known as deoxyribonucleic acid, is a molecule that codes for a host of cellular activities and (in a very real way) determines who you are. DNA strands comprise small coupled units known as nucleotides, which come in four variants

RESEARCH TIP

Distinguish Between 'Genetic' and 'Genealogical' Cousins

What does it mean to be *genetic* cousins, but not *genealogical* cousins? Simply put, DNA establishes a genetic cousin match, while genealogical cousins are relatives who are known through an established paper trail. In addition, you may not share traceable amounts of DNA with more-distant genealogical relatives. Genetic cousins can be genealogical cousins when the paper trail supports them.

and interact to form a double-helix structure in the nucleus of a cell. The four nucleotides—adenine (A), guanine (G), cytosine (C), and thymine (T)—always form in particular ways to create base pairs (specifically, A pairs with T, and C pairs with G), which are important for genetic testing. Human DNA contains about 3 billion base pairs, the vast majority of which are identical amongst all humans.

- DNA is organized into distinct **chromosomes**. Human beings have forty-six chromosomes, organized into twenty-three chromosome pairs. Twenty-two of these are known as **autosomes**, which are the genes tested by most commercial DNA tests (image **A**). The twenty-third pair is made up of the sex chromosomes, which are sex-specific and categorized as either X or Y. Males will have an X chromosome and a Y chromosome (XY), and females have a pair of X chromosomes (XX).

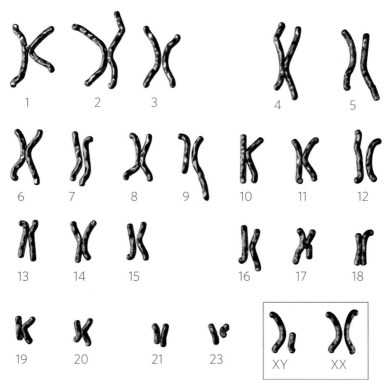

Image A. Humans have forty-six chromosomes. Forty-four of these (called autosomes) are organized into chromosome pairs (called the autosomes), and the final pair makes up the sex chromosomes. Females have two X chromosomes, while males have one X chromosome and one Y chromosome. A person receives half of his chromosomes (one copy of each pair) from each of his parents.

- **Genes** are coded regions on chromosomes that contain specific information. A human has an estimated twenty thousand to twenty-five thousand genes, and you receive one copy of each gene from each parent.

- **Alleles** are multiple forms of a gene that are found on the same area of a chromo-some. You inherit one allele for each trait from each parent, and typically one allele is dominant over the other. One allele on the gene, for example, might code for brown hair, while another allele might be the gene for red hair.

Don't worry—there won't be a test on these terms. But having the right vocabulary is crucial to helping you decide which test to take, as well as how you can interpret and apply your DNA results.

Autosomal DNA Inheritance Patterns

While the underlying biology behind genetics is interesting, you're probably more inter-ested in how it's passed down from generation to generation. By understanding how the different kinds of DNA are inherited, you'll be able to work backwards in time to understand which DNA you received from each parent—and more importantly, how your ancestors are related to you based on your DNA. Most commercial DNA tests deal with **autosomal DNA** (DNA on chromosomes 1 through 22), so we'll look at how that DNA is inherited first. (See chapter 4 for more on how mitochondrial DNA, Y-chromosomal DNA, and X-chromosomal DNA are inherited.)

As we've already stated, you have two copies of each chromosome, one from each of your parents. Your parents, in turn, received 50 percent of their DNA from each of their parents. As a result, you can expect that exactly 50 percent of your DNA came from each parent, roughly 25 percent from each grandparent, roughly 12.5 percent from each great-grandparent, and so on.

However, a couple wrinkles: The autosomal DNA you receive from your parents can undergo a process called **recombination**, in which some of your parent's DNA gets mixed up before being passed on to you. In addition, you inherit different DNA from each parent, meaning your siblings won't share exactly the same DNA with you. You both receive 50 percent of your DNA from each parent, but the DNA making up that 50 percent will be different between the two of you.

I like to illustrate how autosomal DNA transfers from generation to generation using a massive bag of marbles of different colors. Let's say 25 percent of the marbles are red, 25 percent are blue, 25 percent are yellow, and 25 percent are green. Now randomly divide the marbles into three different boxes, then dump them out and repeat the process again and again. Count the marbles in each box, and you'll find a random collection of colors. Though you'll likely have roughly 25 percent of each color in each box, each assortment is unique because the different-colored marbles randomly recombined.

In the same way, DNA inheritance is random. While your father inherited unique DNA from each of his parents (e.g., red and blue marbles), and your mother inherited unique DNA from each of her parents (e.g., green and yellow marbles), the genetic material is never going to be the same as it was when it was passed from your grandparents to your parents, and again when it passed from your parents to you. Each time DNA is inherited, the resulting human being is different.

This helps explain why you share certain amounts of DNA with different genetic relatives, and also why you share *different* DNA with each genetic relative. You receive 50 percent of your DNA from each of your parents and (because your parents received 50 percent of *their* DNA from each of their parents) roughly 25 percent from each of your grandparents. And because DNA is inherited randomly, you share only about 45 to 55 percent of your DNA with full siblings. If your aunt doesn't look much like your mother, this is probably why; she inherited a different set of DNA than did her sister (your mom).

Note that you likely won't be able to tell which autosomal DNA came from which parent. As a result, you won't necessarily know for sure if a match you have on chromosome 1 (that is, someone who shares DNA with you on chromosome 1) is a maternal or paternal relative initially. As a result, testing any known relatives is important to rule out possibilities or confirm new ones.

Let's look at an example. For simplicity's sake, let's say that my father was 50 percent South Asian and 50 percent Eastern European, and my mother was 50 percent Ashkenazi Jewish and 50 percent Irish. They had two children (my brother and me), who will not have exactly the same ethnic breakdown:

	Father's father: 100% South Asian	Father's mother: 100% Eastern European	Mother's father: 100% Ashkenazi Jewish	Mother's mother: 100% Irish
	Father: 50% South Asian, 50% Eastern European		Mother: 50% Ashkenazi Jewish, 50% Irish	
Expected breakdown of each child	25% South Asian	25% Eastern European	25% Ashkenazi Jewish	25% Irish
Reality: Me	23% South Asian	27% Eastern European	24% Ashkenazi Jewish	26% Irish
Reality: My brother	25.5% South Asian	24.5% Eastern European	28% Ashkenazi Jewish	22% Irish

Although both my brother and I would expect to have 25 percent of each ethnicity, our actual results varied from the expectations and from each other. We received different DNA from each parent which made for a mixed bag of "marbles" instead of clean, straight percentages across the generations.

Note that your genetic lines lose more of the older generations' genes as new genetic material is incorporated into your family tree. However, your older ancestors—for genealogical purposes, at least—are the best people to test since they hold more of their ancestors' DNA.

The Testing Process

Now that we've learned the basics of DNA, how do you actually go about taking a test? Once you've selected a test and testing company (see part 2), a DNA test is as simple as spitting in a small vial or swabbing the inside of your cheek with a cotton swab, then putting your sample in a prepaid mailer to send to a lab.

If you're taking a test that requires a saliva sample, be prepared to spit a lot! Only .05 percent of your saliva is made up of cells that comprise your DNA, so you'll need a whole lot of saliva to get a representative sample of DNA. You do have to be able to spit up to the black line—beyond that would be even better. You can stimulate saliva flow either by applying a drop of lemon juice or a pinch of sugar to the tip of your tongue—just enough to cause you to salivate—or by rubbing the cheeks by your back teeth.

If you are using a cotton swab, you will be given two swabs and asked to scrape for thirty seconds on each side of your inner cheek with different swabs to grab enough DNA.

The testing centers will redo the test free of charge if you can't provide enough sample or the test goes missing, but be advised that you always want to create an appropriate sample on your first trial. Having to retake the test multiple times can cost you valuable research time, and (if you are helping someone else submit a test) your test-taker may not be as willing to take the test again—or may not be able to test a second time at all. Fortunately, my ninety-eight-year-old first cousin three times removed was able to provide a second sample after the lab failed to receive her first, but not everyone will be so lucky.

The lab will likely confirm it has received your sample, though it will take longer for it to confirm your sample was large enough to extract data from. The lab then extracts

DNA SECURITY

Some people hesitate to take DNA tests. Some have questions about their privacy, while others think we're leaving ourselves open to being cloned by having that data somewhere available. However, learning about the science and certification should put your mind at ease.

For starters: Commercial DNA tests don't capture enough of your DNA for you to fear being cloned. As discussed in this chapter, testing companies only test 700,000 SNPs out of over ten million. If anyone tries to clone you, they'll get a very incomplete body double!

If that still doesn't suffice, companies have set up numerous safeguards for their customers' privacy. Most companies also keep your samples anonymous, and secure them in a vault with twenty-four-hour surveillance. Finally, the Genetic Information Nondiscrimination Act in the United States makes it illegal for your genetic information to be used by employers or health insurance companies—and how would they get access to that data anyway? Remember, the commercial laboratories we'll discuss in this book are certified and subscribe to a set of best practices that protect consumer privacy.

And even if the test-taker is still not satisfied: He can use an alias or another name entirely to preserve their anonymity.

DNA from the sample and "amplifies" it, essentially creating enough copies of your DNA to run a series of tests on it. Your DNA is placed on a chip, which is then read by a laser on a computer, digitally processed, and verified by data scientists. Results are processed through a server farm of computers that will take your results and compare them with samples tested throughout the world to better learn your ancestry, ethnicity, and (where applicable) health data.

Most tests analyze more than 700,000 regions of your genes, otherwise known as **single nucleotide polymorphisms (SNPs)**. These are individual nucleotides in a genetic sequence that can vary by individual, and comparing these can indicate how closely related test-takers are to each other. The more SNPs where two samples match, the more closely related the two test-takers are. And while the human genome has ten million SNPs and approximately three billion base pairs, 700,000 SNPs are representative enough to draw conclusions.

Limitations of DNA

Although DNA testing is a powerful research tool, it's not foolproof—and you'll need to be careful when drawing conclusions from it. Specifically, some testing companies provide information about your inclination toward specific health conditions, habits, or physical characteristics. Because these tests analyze only certain segments of your DNA, they can't capture your whole genetic profile. As a result, the testing companies can't predict these characteristics with as much certainty as they would have you believe.

For example, on chromosome 22, marker rs6269 has been linked to pain sensitivity, while four markers on chromosome 15 are said to address smoking behavior, caffeine consumption, caffeine metabolism, and eye color. However, the traits associated with these genetic markers don't necessarily determine your relationship with pain, smoking, or caffeine. Rather, they just tell you what (compared with others) you're genetically predisposed to. Case in point: I'm a nonsmoker with the AG base-pair on rs1051730, which indicates I likely smoke every day. However, I haven't smoked a single cigarette in my life. I suppose that, should I choose to smoke, my genetics might predispose me to becoming addicted to it, but my personal choices have invalidated that genetic tendency.

Even if these SNPs *could* consistently predict behavior or physical traits, other factors could invalidate them. The presence of another marker may negate one SNP's coding, for example, or perhaps environmental factors offset the effect of an SNP. It's also possible that we don't fully understand the connection between genetic markers and a particular trait, making it difficult to predict.

Here's another example that puts some of DNA's limitations on display. Image **B** shows the Predict Eye Color tool on the website GEDmatch **<www.gedmatch.com>** (more on this

CC at: rs3794604 - Blocks some melanin. Often gives light colored eyes.
GG at: rs7174027 - Blocks some melanin. Often gives light colored eyes.
CC at: rs4778241 - Low Melanin. Basis for Gray, Blue, Green, or Yellow Eyes if no other pigmentation is present.
CC at: rs9782955 - Blocks some melanin. Often gives light colored eyes.
CT at: rs3947367 - Contrasting sphincter around pupil.
CC at: rs1129038 - High Melanin production. Brown.
AA at: rs1105879 - Weak Amber Gradient
AG at: rs10467971 - Penetrance Modifier - Blue
GT at: rs7713279 - Inhibit weak amber gradient
GG at: rs12906280 - Gray ring around outer edge
TT at: rs35405 - Inhibit weak amber gradient

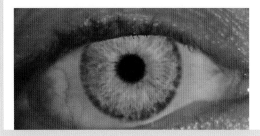

Image B. GEDmatch is a service that does pretty nifty things with DNA results, including estimating your eye color (though it's not always accurate). Those letters followed by numbers, like rs35405, are the actual SNPs used for analysis.

in chapter 10). According to it, my eye color should be a beautiful blue based on my genes. I've always wanted blue eyes, but mine are as brown as can be.

You can derive several theories from your base-pair combos, but they're not to be taken for fact. Health-related results (such as predispositions to certain conditions) can be especially scary, but don't consume yourself with data that may not be true. It's fun to find out a little about what your DNA might say about you, but it certainly can cause false fear. The more important task for adoptees is discovering the people you match with, not what determining what your DNA tells you about your health or physical appearance.

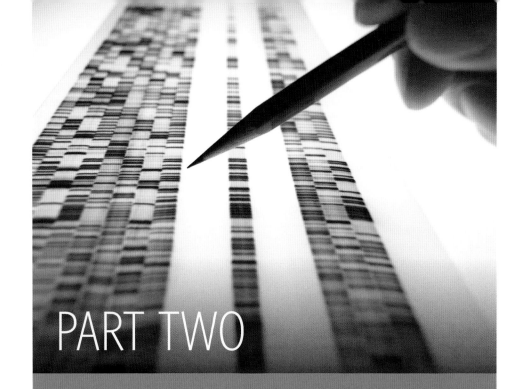

PART TWO

DNA Tests and Testing Companies

4

Types of DNA

n chapter 3, we discussed the basics of DNA testing and its underlying biology. As you learned, submitting a sample is painless and straightforward, and DNA can help resolve some of your trickiest research problems. But what do all the different kinds of DNA do? And which should you use in your research?

In this chapter, we'll discuss the four main types of DNA used in genealogical testing: autosomal DNA, X-chromosomal DNA (X-DNA), mitochondrial DNA (mtDNA), and Y-chromosomal DNA (Y-DNA). We'll also discuss the practical uses of each kind of DNA and briefly review the science behind each test so you know what to expect and how to best understand your results.

Note that some testing companies (such as AncestryDNA **<dna.ancestry.com>** and MyHeritage DNA **<www.myheritage.com/dna>**) test only for autosomal DNA, which is generally the most useful for genealogical problems. Family Tree DNA **<www.familytreedna.com>** also tests for autosomal DNA, and in addition offers mtDNA and Y-DNA testing, which are crucial data pieces for specific research queries. 23andMe **<www.23andme.com>** is a bit of an odd bird—it offers only an autosomal DNA test, but results include mtDNA and Y-DNA haplogroup information like a test dedicated to each respective kind of DNA would.

You're usually better off taking an autosomal DNA test, then moving onto mtDNA or Y-DNA in specific cases. (X-DNA is often tested as part of your autosomal DNA test, so you likely won't have to test this separately.)

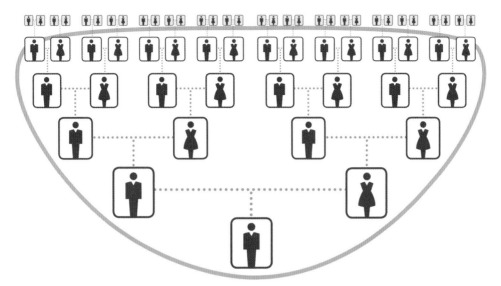

Image A. Autosomal DNA looks at all your relatives from the last several generations, as indicated by this circle. When you do an autosomal test for genealogical purposes, you always want to test the older generation if possible. But if you're looking for a parent, sibling, or child, go with your gut and test the closest.

Autosomal DNA

Autosomal DNA is the DNA inherited on the twenty-two chromosome pairs that both male and females have. (The twenty-third chromosome determines sex and carries X- or Y-DNA, and we'll discuss these in the respective sections later.) As previously mentioned, autosomal DNA will generally give you the highest chance of reconnecting with biological family members because you have received DNA from recent ancestors (image **A**).

In chapter 3, we discussed some of the basics behind autosomal DNA and how it's passed down from one generation to the next. You receive 50 percent of your autosomal DNA from each of your parents, though which DNA segments you receive will vary. In addition, autosomal DNA recombines during the reproductive process, meaning some genetic information is spontaneously changed, leading to more discrepancies between your and your parents' DNA.

Autosomal DNA tests measure the amount of autosomal DNA you share with other test-takers. Shared DNA is measured in units called centimorgans (cM), and the testing services will suggest genetic matches to you based on the number of cM you share with another test-taker. As you might expect, you're more likely to have a close relationship with someone who shares higher amounts of DNA with you. For example, a child typically shares close to 3,600 cM with a parent, meaning that a relationship with that amount of shared DNA is likely parent-child. Likewise, full siblings share an average of 2,650 cM, while half-siblings share about 1,800 cM.

In this way, you can find relatives by using the amount of shared cM along with the chart in image **B** to determine what relationship you and a genetic match might be—or sister, aunt, brother, half-sister, half-aunt, etc. This chart is a great reference for determining relationships, and you'll find that it's used in many DNA circles. Note that different testing services have different ways of calculating cM (as well as different shared-cM thresholds for matching you with other users), but we'll discuss this in more detail in chapter 5.

DNA Detectives Autosomal Statistics Chart						
Created by Christa Stalcup						©THEDNADETECTIVES, 2016
cM (centiMorgans)^		**Percentage (%) of Shared DNA^^**		**Group**	**Relationship**	**Notes**
Average	**Range**	*Average*	**Range**			
3,600		50%			Parent - Child	
2650	2300 - 3900	37%	32%-54%	Group A	Full Sibling	Ancestry, FTDNA and GEDmatch (HIR only)
3600		50%				23andMe (FIR included)
1800	1300 – 2300	25%	18%-32%	Group B	Half Sibling Aunt/Uncle/Niece/Nephew Double First Cousin Grandparent/Grandchild	3/4 Siblings^^^
900	575 - 1330	12.5%	8%-18.5%	Group C	First Cousin (1C) Half Aunt/Uncle/Niece/Nephew Great-Grandparent/Great-Grandchild Great-Aunt/Uncle/Niece/Nephew	
450	215 - 650	6.25%	3%-9%	Group D	First Cousin Once Removed (1C1R) Half First Cousin (½ 1C) Half Great-Aunt/Uncle/Niece/Nephew	
224	75 - 360	3.125%	1%-5%	Group E	Second Cousin (2C) First Cousin Twice Removed (1C2R) Half First Cousin Once Removed (½1C1R)	
112	30 - 215	1.56%	0.42% - 3%	Group F	Second Cousin Once Removed (2C1R) Half Second Cousin (½ 2C) First Cousin Three Times Removed (1C3R) Half First Cousin Twice Removed (½ 1C2R)	
56	0 - 109*	0.78%	0% - 1.52%	Group G	Third Cousin (3C) Second Cousin Twice Removed (2C2R)	~10% of 3Cs will not share DNA*
30	0 - 75**	0.4%	0%-1%	Group H	Third Cousin Once Removed (3C1R) Other Distant Cousins	~50% of 4Cs will not share DNA**

^cM =Ancestry.com & FTDNA
^^Percentage of DNA = 23AndMe
^^^ 3/4 Siblings are a combination of half siblings and 1ˢᵗ cousins, FIRs are included.

Groups A & B: 99% within the ranges given
Groups C – I: 95% within the ranges given

Image B. The DNA Detectives group has put together this helpful chart, which is probably the most important chart you'll use in your research. It will help you determine your relationship to another person based on the amount of DNA you share.

It may be more difficult to find birth family members further down the cM relationship spectrum, especially since you may not have the benefit of relatives' knowledge or cooperation. For example, when researching the parentage of an adoptee in my family, I connected with a second cousin who knew the birth family of two sibling adoptees very well, but had no idea they had been given up for adoption. This often happens, and the family you reach out to may be in denial.

A second chart, created in August 2017 by Blaine T. Bettinger **<thegeneticgenealogist. com/wp-content/uploads/2017/08/Shared_cM_Project_2017.pdf>** (image **C**), shows similar findings. Using a sample size of more than twenty-five thousand participants, Bettinger estimates cM ranges for various relationships, categorized in Clusters (just as the green chart from DNA Detectives uses groups). He also adds columns for accuracy ranges, helping you know at what levels you can assert a relationship at the 95th and 99th accuracy percentiles. You can find interactive versions online **<dnapainter.com/tools>**,

FULL VERSUS HALF-IDENTICAL REGIONS

So you match on a chromosome with another individual. But do both of your alleles match with that person, or just one? More-advanced genetic genealogists use two classifications, full identical regions (FIRs) and half-identical regions (HIRs) to distinguish between these two ways of sharing DNA.

An FIR means just what you think it does: A region of DNA matches on both alleles. To put it another way: You and your fellow test-taker share DNA on both maternal and paternal lines at an FIR. Generally, only full siblings and double cousins will have many FIRs.

Likewise, an HIR is a region of a chromosome pair in which one of the alleles differ. To put it another way: An HIR occurs when you and a test-taker share DNA on a maternal or paternal line at a particular region, but not both. For example, you share an HIR with your paternal first cousin because the match is only paternal, not maternal.

This concept can be a bit tricky, so let's look at an example. If I inherited a DNA segment of TTTTTT from my father and GGGGGG from my mother, anyone who matches at TGGTTT or GGTGTG would appear as a match because we shared segments of DNA. However, that would be an HIR match. A FIR would only be someone who matches exactly with TTTTTT or a GGGGGG.

Cluster	Relationships	Total #	Average	Range (95th Percentile)	Range (99th Percentile)	Expected
Cluster #1	Siblings	1345	2629	2342 - 2917	2209 – 3384	2550
Cluster #2	Half Sibling, Aunt/Uncle/Niece/Nephew, and Grandparent/Grandchild	2473	1760	1435 – 2083	1294 – 2230	1700
Cluster #3	1C, Half Aunt/Uncle/Niece/Nephew, Great-Grandparent/Great-Grandchild, and Great-Aunt/Uncle/Niece/Nephew	2261	884	619 – 1159	486 – 1761	850
Cluster #4	1C1R, Half 1C, Half Great-Aunt/Uncle/Niece/Nephew, and Great-Great Aunt/Uncle/Niece/Nephew	1842	440	235 – 665	131 – 851	425
Cluster #5	1C2R, Half 1C1R, 2C, and Half Great-Great-Aunt/Uncle/Niece/Nephew	2224	232	99 – 397	47 – 517	213
Cluster #6	1C3R, Half 1C2R, Half 2C, and 2C1R	2284	123	0 – 236	0 – 317	106
Cluster #7	Half 1C3R, Half 2C1R, 2C2R, and 3C	2492	75	0 – 158	0 – 229	53
Cluster #8	Half 2C2R, 2C3R, Half 3C, and 3C1R	1864	49	0 – 114	0 – 175	27
Cluster #9	Half 3C1R, 3C2R, and 4C	1528	36	0 – 84	0 – 122	13
Cluster #10	Half 3C2R, 3C3R, Half 4C, and 4C1R	1040	29	0 – 67	0 – 118	7

The Shared cM Project – Version 3.0
August 2017

Blaine T. Bettinger
www.TheGeneticGenealogist.com
CC 4.0 Attribution License

For MUCH more information (including histograms and company breakdowns) see: goo.gl/Z1EcJQ

Image C. Like the green chart by DNA Detectives, this chart groups known relationships into clusters based on the total number of test-takers, the average among each person, ranges at 95 and 99 percent confidence, as well as the expected number of shared cM between each relationship cluster. This is the other most important chart you'll use in your search.

created by Jonny Perl. Bettinger also created a version of the chart to showcase genealogical (as well as genetic) relationships (image **D**).

As with all things in genetic genealogy, shared cM values are useful tools, but not absolute. Depending on your family's circumstances, you may observe slightly different amounts of shared DNA for the same relationships—see the Endogamous Cultures and Shared cM sidebar for an example. You should also consider on what chromosome(s) you share DNA with another test-taker. Do you have long segments of shared DNA (15 cM or higher for non-endogamous communities, 20 cM or higher for endogamous families), or is your shared DNA spread out across multiple chromosomes (e.g., less than 7-cM segments)? If the latter, you may have more trouble understanding the nature of your relationship with that other test-taker. Not all services provide this level of detail, so in all likelihood we're going to have to use a third-party service to tell you what to look for; we'll discuss that in chapter 10.

The Shared cM Project – Version 3.0
August 2017

Blaine T. Bettinger
www.TheGeneticGenealogist.com
CC 4.0 Attribution License

For MUCH more information (including histograms and company breakdowns) see: goo.gl/Z1EcJQ

How to read this chart:

Aunt/Uncle	→ Relationship
1750	→ Average
1349 – 2175	→ Range (low–high) (99% Percentile)

Relationship chart (each cell: Relationship — Average (Range low–high)):

	Half GG-A/U line	Half G-A/U line	Half A/U line	Half-Sibling line	Direct line	Sibling line	1C line	2C line	3C line	4C line	5C line
g5					Great-Great-Great-Grandparent						GGGG-Aunt/Uncle
g4					Great-Great-Grandparent					GGG-Aunt/Uncle	
g3	Half GG-Aunt/Uncle 187 (12–383)				Great-Grandparent 881 (464–1486)				Great-Great Aunt/Uncle 427 (191–885)		
g2		Half Great-Aunt/Uncle 432 (125–765)			Grandparent 1766 (1156–2311)			Great Aunt/Uncle 914 (251–2108)			
g1			Half Aunt/Uncle 891 (500–1446)		Parent 3487 (3330–3720)		Aunt/Uncle 1750 (1349–2175)				
g0 (self)	Half 3c 61 (0–178)	Half 2c 117 (9–397)	Half 1C 457 (137–856)	Half-Sibling 1783 (1317–2312)	SELF	Sibling 2629 (2209–3384)	1C 874 (553–1225)	2c 233 (46–515)	3c 74 (0–217)	4c 35 (0–127)	5c 25 (0–94)
g-1 (child)	Half 3c1R 42 (0–165)	Half 2c1R 73 (0–341)	Half 1C1R 226 (57–530)	Half Niece/Nephew 891 (500–1446)	Child 3487 (3330–3720)	Niece/Nephew 1750 (1349–2175)	1C1R 439 (141–851)	2c1R 123 (0–316)	3c1R 48 (0–173)	4c1R 28 (0–117)	5c1R 21 (0–79)
g-2 (grandchild)	Half 3c2R 34 (0–96)	Half 2c2R 61 (0–353)	Half 1C2R 145 (37–360)	Half Great Niece/Nephew 432 (125–765)	Grandchild 1766 (1156–2311)	Great-Niece/Nephew 910 (251–2108)	1C2R 229 (43–531)	2c2R 74 (0–261)	3c2R 35 (0–116)	4c2R 22 (0–109)	5c2R 17 (0–43)
g-3 (great-grandchild)	Half 3c3R	Half 2c3R	Half 1C3R 87 (0–191)	Half GG Niece/Nephew 187 (12–383)	Great-Grandchild 881 (464–1486)	Great-Great-Niece/Nephew 427 (191–885)	1C3R 123 (0–283)	2c3R 57 (0–139)	3c3R 22 (0–69)	4c3R 29 (0–82)	5c3R 11 (0–44)

Other Relationships

Relationship	Average	Range
6C	21	0 – 86
6C1R	16	0 – 72
6C2R	17	0 – 75
7C	13	0 – 57
7C1R	13	0 – 53
8C	12	0 – 50

Minimum was automatically set to 0 cM for relationships more distant than Half 2C, and averages were determined only for submissions in which DNA was shared.

Image D. This chart, also created by Blaine T. Bettinger, calculates the average amount of shared cM between various relationships and plots them out in a relationship chart.

X-DNA

When you take an autosomal DNA test, you're likely testing more than just your autosomal DNA. These tests also look at your X chromosome, another category of genetic information that can contain valuable information about your family. X-DNA, as the name implies, is genetic information that lies on the X chromosome, one of the two chromosomes that determine biological sex. Recall that males have one X chromosome (given by the mother), while females have two X chromosomes, one from each parent. Note that you'll likely need third-party tools to help you view your X-DNA data, as some companies (such as AncestryDNA) don't show X-DNA apart from your autosomal DNA results.

Like all DNA, you'll find there are pros and cons to using X-DNA. The good news: Because of how X-DNA is inherited, you may be able to make concrete determinations about how you and an X-DNA match share DNA. However, these inheritance patterns can be tricky, as they dictate you can only receive X-DNA from specific genetic lines.

If you have a male X-DNA match, for example, you know you can trace his X-DNA back to his mother's side, then make inferences from there. As seen in image **E** (which shows how X-DNA is inherited down the generations for a male test-taker), males always inherit one of their mother's X chromosomes. His mother received that X chromosome from either her father or her mother, who received it from one of their parents (and so on).

Tracking X-DNA down a female line, however, is even trickier. Image **F** shows how a female test-taker received her X-DNA. Since she received an X chromosome from each parent, you might have trouble determining on which branch of your family tree you're an X-DNA match with someone. As we already know, the X chromosome she received from

ENDOGAMOUS CULTURES AND SHARED cM

I f your population is endogamous (that is, a culture in which marriages occurred within the same tribal group for generations), you'll likely observe different amounts of shared cM between genetic relatives than is typical. For example, third cousins typically share 0 to 109 cM, but third cousins in endogamous communities may share much more DNA than that because they're related to other test-takers in more than one way. It is recommended by the International Society of Genetic Genealogy that you find a single segment that is at least 23 cM to find relatives. See chapter 3 for more on how endogamy can affect your DNA research.

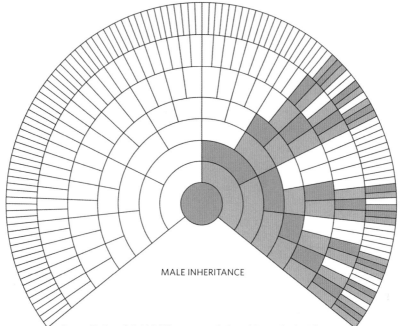

MALE INHERITANCE

Image E. A male's X-DNA comes only from his mother's side, which is often helpful in establishing matches—especially since he knows only to look up his mother's line.

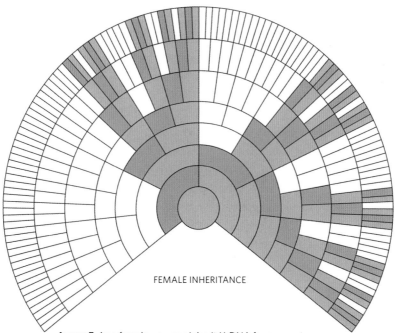

FEMALE INHERITANCE

Image F. As a female, you can inherit X-DNA from ancestors on both sides of your family, making it more difficult to determine which X-DNA line to follow up the generations.

her father had to come from *his* mother (the test-taker's paternal grandmother), who, in turn, got her X chromosomes from both of her parents. Her mother received X chromosomes from both of her parents, too.

Let's look at this more closely, using how you can inherit X-DNA from great-grandparents as an example. X-DNA inheritance patterns dictate that male and female test-takers can inherit their X-DNA from different great-grandparents. A male can inherit X-DNA from just three great-grandparents: his maternal grandfather's mother, his maternal grandmother's mother, and his maternal grandmother's father. A female, however, can inherit X-DNA from five great-grandparents: her paternal grandmother's father, her paternal grandmother's mother, her maternal grandfather's mother, her maternal grandmother's father, and her maternal grandmother's mother. To sum up: Men researching X-DNA only need to worry about their maternal side, but women need to be mindful of both maternal and paternal sides.

What does this all mean? If you have an X-DNA match, you should carefully consider which sex the test-taker is, as this will affect what X-DNA he or she inherited (and, thus, how you might be related). For example, if you have two males who tested their X-DNA and match, their shared relatives must be on each of their maternal sides (as men always receive X-DNA from their mother's family). You can use this as a clue to look down each test-taker's maternal line for a shared ancestor.

One other caveat to X-DNA testing: X-DNA doesn't recombine as often as autosomal DNA, so (depending on which X-DNA you share with someone) your genealogical relationship may go back further in time than you can genealogically prove. Therefore, use X-DNA in conjunction with other kinds of data (such as shared autosomal DNA) before you put too much stock into a potential X-DNA match.

mtDNA

As we learned in chapter 3, one key type of genetic material is found outside of the cell's nucleus: mtDNA. Found in the mitochondria, the "powerhouse" of the cell that enables the cell to produce energy, mtDNA can give us insight about a specific line of our research. Mitochondria (and mtDNA) are passed down from mother to child, meaning (whether you're male or female) you received your mtDNA from your mother, who received it from her mother, who received it from her mother, and so on since the beginning of human existence. As a result, mtDNA can help you trace your mother's mother's mother's (etc.) line. (Note: This means that an mtDNA line "dies" with the male. A male can test his mtDNA, but it will not pass to his son or his daughter. Instead, the mtDNA of his children will come from their mother, which is likely a different mtDNA line entirely.)

As a result, mtDNA is great for finding genetic cousins along your strictly maternal line. Image **G** shows an example of how mtDNA is inherited throughout the generations. Note that siblings all share mtDNA with each other and with their mother, but only the female siblings will pass their mtDNA down to their children. Image **H** shows a more direct summary of which ancestors you can expect to learn about using mtDNA.

If you decide to take an mtDNA test, your results will come in the form of a list of mutations, plus a haplogroup assignment. Haplogroups are groups who share an ancestral

mtDNA Inheritance Descendants Chart (Maternal Line)

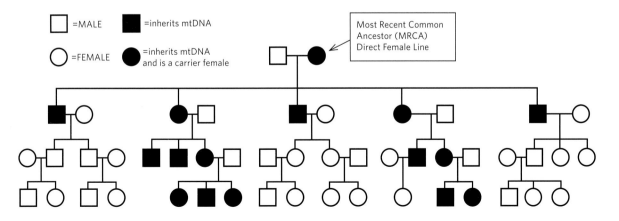

Image G. mtDNA can only be inherited from mothers. This image shows mtDNA can be shared across generations, with the black circles and squares representing people who will have the same mtDNA.

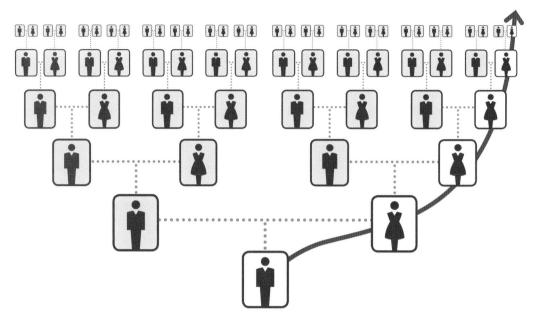

Image H. Mothers pass their mtDNA down to both of their children. So when taking an mtDNA test, you'll learn about your directly maternal lines.

origin and demonstrate certain genetic markers, identified by researchers. Haplogroups are determined by identifying clusters of alleles in tightly linked genes that are characteristic of a specific people group thousands of years ago. Most (but not all) haplogroups have been identified alongside a specific location of origin.

mtDNA haplogroups are usually defined by alphanumeric characters, and all haplogroups are related if you research long enough back in time. The haplotype L represents the first human from whom all of us genetically descend—we call her Mitochondrial Eve, and she resided in Africa between 151,000 and 233,000 years ago. Mitochondrial Eve is believed to be the most recent common ancestor for all of humankind, but it is possible there were other matriarchs who died off before her.

The L haplogroup, then, is the root of the evolutionary tree. As the generations followed from Mitochondrial Eve, subclades (subgroups of a haplogroup created by mutations) were formed. These were called L0, L1, L2, L3, L4, L5, and L6, and most non-Africans descended from haplogroup L3. Haplogroup L3 further mutated into haplogroups M and N. As the population distribution became spread out and humans migrated to newer areas, newer haplogroups were formed through genetic mutations.

Phylogenetic tree of human mitochondrial DNA (mtDNA) haplogroups

Mitochondrial Eve (L)

L0					L1–6								
	L1 L2				L3						L4 L5 L6		
		M				N							
	CZ D E G Q			O A S		R			I W X Y				
	C Z					B F R0 pre-JT P U							
						HV JT K						
						H V J T						

Image I. All mtDNA haplogroups are related. This family tree of sorts shows how modern haplogroups descend from earlier haplogroups, starting from Mitochondrial Eve (also known as haplogroup L).

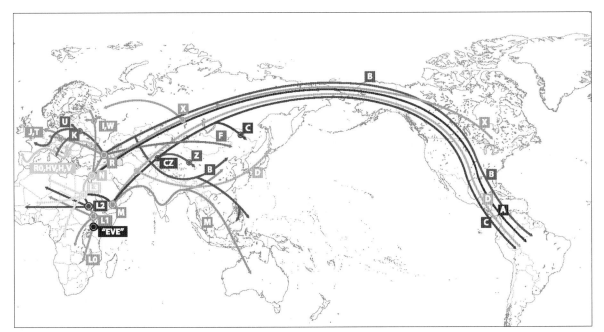

Image J. Archeologists and genetic historians have determined how the various haplogroups migrated over the millennia.

Genetic genealogists have studied these rates of mutation and calculated the times in which these various people groups originated. The study of phylogenetics (the evolutionary development and diversification of species) has resulted in a phylogenetic mtDNA tree (image **I**). We've already established that L3 has mutated into M and N. As you can see, CZ, D, E, G, and Q emerged from M, and N drills down even further. Over evolutionary time, more and more mutations will be discovered both from our past and our future. Image **J** maps out how haplogroups have migrated, splintered, and evolved over the millennia.

In addition to giving you a haplotype, your mtDNA results (available only on Family Tree DNA <www.familytreedna.com>) will show you a genetic distance. A genetic distance of zero means that you share the exact same mitochondrial DNA as your match, while a genetic distance of one (or five, or however many) shows that there is one (or five, etc.) mutations between you and other people in your haplogroup.

Let's look at an example. Image **K** shows my mtDNA matches, organized by genetic distance. With further research, I determined none of my close mtDNA matches (those who have a genetic distance of zero) also share large amounts of autosomal DNA with me. I've conversed with a few of them, and one seems to have similar origins to mine—but even she's not close enough to make a clearly established match. If I can find another relative to also test or access records that tie us together, I may be able to bridge that gap—but it's unlikely that we'll make a connection in the near future.

HVR1, HVR2, CODING REGIONS - 68 MATCHES					
					Page: 1 2 3 of 3
Genetic Distance	Name			Earliest Known Ancestor	mtDNA Haplogroup	Match Date
0		FMS FF		Pauline ▓▓▓▓ 1883-1972	U4a3a	1/17/2017
0		FMS FF			U4a3a	1/9/2017
0		FMS		Sarah ▓▓▓▓ b.1876 d 1952	U4a3a	12/27/2016
0		FMS FF			U4a3a	12/27/2016
1		FMS FF			U4a3a	9/29/2017
1		FMS FF		Callie ▓▓▓ b January 21, 1872 and d Aug 7, 1905	U4a3a	2/3/2017
1		FMS FF			U4a3a	2/3/2017
1		FMS FF			U4a3a	12/27/2016
1		FMS FF		Franticek ▓▓▓ b before 1780	U4a3a	12/27/2016
1		FMS FF		Mary Magdelene ▓▓▓ b.1790 d.1829 Switzerland	U4a3a	12/27/2016
1		FMS FF		Eliza (Unknown) ▓▓▓, b. 1818 d. 1896	U4a3a	12/27/2016

Image K. Genetic distance indicates how your mtDNA compares with another test-taker's, with each genetic distance unit representing a variation between the two tests.

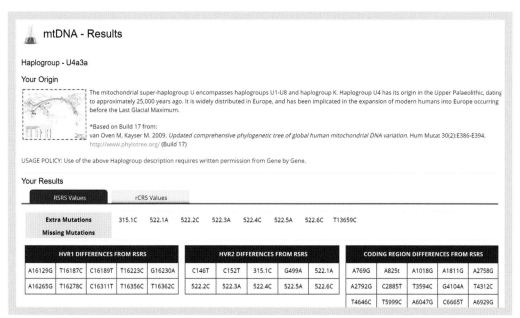

Image L. You'll view your mtDNA results as a list of differences from the "RSRS" (Reconstructed Sapiens Reference Sequence), which is a constructed human genome used when comparing mtDNA.

Those more experienced with DNA should take a look at image **L**, which displays my raw results. Here, you'll see a list of variations in my mtDNA as compared to the RSRS (Reconstructed Sapiens Reference Sequence), a standard that researchers have created to help identify mtDNA variations.

Now that we've covered the basics of mtDNA, you should make a couple quick notes before signing up for an mtDNA test. First, mtDNA mutates much more slowly than other kinds of DNA. As a result, your mtDNA likely hasn't changed much throughout the generations, meaning it can be difficult to use it to distinguish between relatives (e.g., your mother from your maternal aunt). Similarly, your mtDNA haplogroup only tells you about your "deep ancestry." While it can help you identify other haplogroup members/mtDNA relatives, mtDNA won't help you determine how you're related to these individuals.

For example, I am in haplogroup U4a3a (not just haplogroup U), meaning other researchers in that same haplogroup may have recent ancestry with me. U4 dates back fifteen to twenty-five thousand years and originates in Central Asia, and U4 mutated into U4a3 about four thousand years ago.

Though all the members of my haplogroup were closely related way back then, we may not share a common ancestor more recently than a couple thousand years ago. While

I may find matches who are genetically related to me, I can't prove the connections without a true paper trail, which is hard to document after more than a couple hundred years. mtDNA can be useful in proving close relationships between strictly maternal cousins and siblings, but (without other tests, such as autosomal DNA), it is not particularly useful in finding close relatives.

mtDNA, then, doesn't have as much staying power as autosomal DNA, and I wouldn't recommend taking it unless you have a specific research goal in mind. For example, mtDNA is great at determining if you share an mtDNA haplogroup with someone else (say, an autosomal DNA match with whom you're trying to establish a relationship) to ascertain maternal lines. mtDNA tests can also help you determine if a match is maternal or paternal, which can help you narrow down your research. For example, an adoptee named Steven was able to determine that a second-cousin autosomal DNA match wasn't a direct maternal relative of his mother. She didn't match the second cousin on the mtDNA line, even though the paper trail seemed to indicate she should. Rather, he and a fellow researcher assume his great-grandfather may have had a first wife (not documented) from whom he descends. Based on the data the families and researchers have available, he may be right.

Y-DNA

Now that we've established the significance of mtDNA, we have one last kind of DNA to cover: Y-DNA. As we discussed in chapter 3, only males have a Y chromosome, and they always receive the Y chromosome from their fathers. As a result, Y-DNA is passed from male to male to male to male on a direct paternal line. Much like mtDNA allows you to research your strictly maternal line, Y-DNA allows you to examine your strictly paternal line: your father, your father's father, your father's father's father, etc.

Let's look at an example of how Y-DNA is inherited in a family. If a male has only daughters, he will not pass on any Y-DNA, and the Y-DNA line dies there with the father (unless, of course, he had a brother who had sons). If the males have sons, those sons will inherit the Y-DNA and pass it onto their sons and grandsons, and so on. Image **M** shows how Y-DNA appears in a particular family.

Still confused? Check out image **N** which displays the path of Y-DNA throughout the generations. Of course, non-highlighted males in this image still possess Y-DNA, but it's different Y-DNA from the line we're researching.

Like mtDNA, researchers have identified Y-DNA haplogroups that trace the estimated origins of the paternal line. When you take a Y-DNA test, you'll be placed into one of these haplogroups, and you can use this to estimate your ancestral origins and group yourself

Y-DNA Inheritance Descendants Chart (Paternal Line)

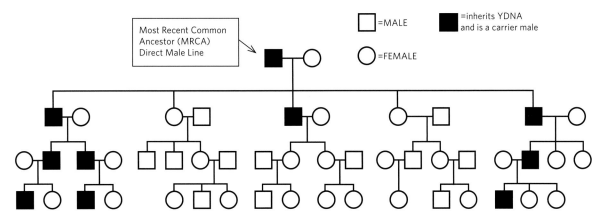

Image M. Like mtDNA, Y-DNA has a specific inheritance pattern: Only men have Y-DNA, and fathers pass on their Y-DNA only to sons. In this chart, the black squares and circles indicate relatives who have the same Y-DNA.

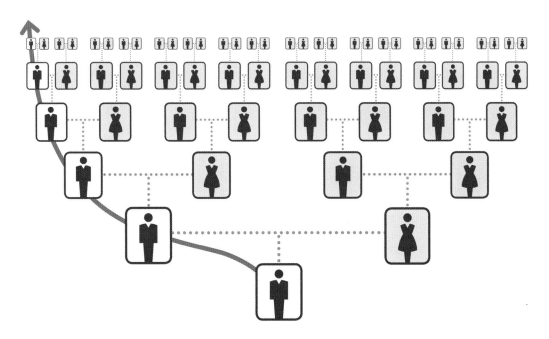

Image N. This easy-to-understand visual shows the general direction that your Y-DNA research will take.

with other Y-DNA test-takers. However, Y-DNA recombines more rapidly than mtDNA (at a rate of two mutations per generation), resulting in each haplogroup containing hundreds to thousands of mutations.

Image **O** shows the history of Y-DNA haplogroups throughout time. And as with mtDNA (which has a calculated Mitochondrial Eve), researchers have identified a "Y-chromosomal Adam," the patrilineal most recent common ancestor of all humans. Origins of the date of Y-chromosomal Adam are unknown, though an estimate places the first haplogroup, A00, in Africa some 235,900 years ago. The haplogroups that have since emerged from Y-chromosomal Adam are estimated to have originated about sixty-five thousand years ago, beginning with E.

Y-DNA is a bit complicated because people elect to take different tests, and the results are more difficult to interpret than results from other tests. At the time of this printing, Family Tree DNA offers 37-, 67-, or 111-marker Y-DNA tests. (A bigger test, the Big Y, will be explained shortly.) All of these Y-DNA tests are short tandem repeat (STR) tests that evaluate multiple copies of repeated sequences of nucleotide bases. For example, if your DNA repeats the sequence TAGA ten times at a specific marker, you will have a value of 10 at that region. In general, those who match on a high number of markers are more closely

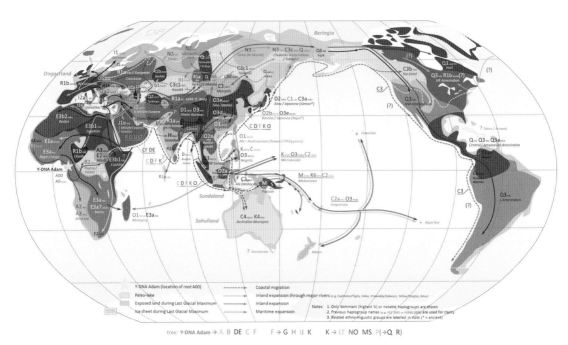

Image O. Y-DNA haplogroups all descend from a common "Y-Chromosomal Adam," and historians have been able to trace the migrations of Y-DNA haplogroups.

related. The more markers you test, the more accurate your results will be—but you'll have fewer matches as a result, since not everyone tests all the markers.

Comparing results across these multiple kinds of tests can skew your data, and you may even miss some potential genetic matches by viewing them only in one way. For example, my father took a Y-67 test, and his paternal second cousin, Gary, took the less-extensive Y-12 test. My father's default screen is to view results at sixty-seven testing sites, and this level shows he has no one at the genetic distance of 0 on the Y-67 marker. However, if he views his results from fewer markers (say, Y-12), my father can view several matches at a genetic distance of 0—including Gary. This is further complicated by the fact that Family Tree DNA states there's only a 33.57 percent chance that Gary and my father have shared an ancestor in the past four generations.

As a result, test with as many markers as possible to open yourself up to the most possibilities and ways of analyzing your data. Doing so will enable you to review results at various marker thresholds, providing you with the greatest chance of finding a genetic relative. Tests with more markers also give you more detailed haplogroups. My father took the most comprehensive Y-DNA test available (Big Y), for example, and his haplogroup is more specific than Gary's.

For adoptees, Y-DNA tests can be excellent tools in determining paternity. Specifically, Y-DNA tests can lead you to a birth surname and a haplogroup, either of which can connect you to genetic cousins or even closer relatives. Your results may even give you a paternal country of origin, the total number of known matches for that specific country, the total number of people in the country who have reported a match (but who may not be a Y-DNA match to you), the percentage from the country of origin compared with the total number within Family Tree DNA's database, and any additional comments provided by other test-takers.

5

AncestryDNA

N ow that we've gotten all the science out of the way, let's look at the different tests and services available. Some companies offer different tests than others, and even the same kinds of DNA tests may be administered or processed differently at the various companies. The following chapters will give you the lowdown on each major testing service, beginning with AncestryDNA **<dna.ancestry.com>**.

By far, Ancestry comprises the largest DNA database around. As of this book's publication, more than seven million people have tested with AncestryDNA—all of them autosomal DNA. That means that you likely have the best chance of finding close cousins, but you will not be able to get any maternal or paternal haplogroups.

Testing with AncestryDNA is relatively easy. Like the tests we discussed in chapter 3, you only need to spit into a tube to provide your DNA sample. To get your test ready, you will need to activate the kit using the alphanumeric characters that are unique to every test and will identify your test in the lab. Recall in chapter 3 that if you have trouble

RESEARCH TIP

Watch for Changes

Because services occasionally tend to change their design layouts, the screenshots you see in this chapter and the chapters that follow may not match what you see on the screen. Despite this, the underlying principles and strategies will remain the same. And since we're still just at the beginning of the age of genetic genealogy, you can also expect newer features to launch that make DNA analysis even easier.

spitting, you can use a pinch of sugar or lemon juice to help you salivate. Once you send off your sample, the company will notify you when the package was received, when it went to the lab for processing, and when your results are ready. Depending on how busy the season is, results can come in two weeks or twelve. Expect response times during the holidays to be particularly lengthy.

Your results from AncestryDNA will give you two core insights: your ethnicity estimate and your DNA matches. If you're looking for biological family, you're going to want to take a look at your DNA matches first. However, the ethnicity insights Ancestry provides can have expected or unexpected ramifications. Let's look at each in turn.

Ethnicity Estimates

Ethnicity estimates have earned a reputation in the genealogical community for being imprecise—guesses at ethnic origins, rather than facts. But they can provide key evidence for uncovering family secrets. In a *Washington Post* article published in June 2017 **<www.washingtonpost.com/graphics/2017/lifestyle/she-thought-she-was-irish-until-a-dna-test-opened-a-100-year-old-mystery>**, Alice Collins Plebuch, who believed herself to be Irish, found out that her father was European Jewish. With such deeply embedded roots in the

DNA AND YOUR ANCESTRY.COM FAMILY TREE

Many of AncestryDNA's analysis tools rely on you uploading a family tree to Ancestry.com, then connecting your DNA results to a person within the tree. To make the most out of AncestryDNA's test, be sure to do this, as it will make your search for genetic relatives much easier and provide you with more robust analysis tools.

Here's how. You'll first need to create a family tree, which you can do at **<www.ancestry.com/family-tree>**. (See chapter 9 for how to build a family tree if you don't yet know your birth family). Note that you can make your family tree private for now, if that makes you more comfortable. You may not want others to know you're looking for immediate family members—at least, not yet. To connect your DNA results, go to your DNA results page and click Settings. Under the Family Tree Linking section, add yourself to your family tree.

When we talk more about mirror trees in chapter 9, you'll know the importance of linking your DNA results to a tree of relatives as determined by close DNA matches. As you build out your family tree, this will allow the system to find common ancestors between you and your shared matches.

For more on uploading and maintaining a family tree on Ancestry.com, check out the *Unofficial Guide to Ancestry.com* by Nancy Hendrickson (Family Tree Books, 2018).

Irish community, she was confused and distraught that her genetic ethnicity wasn't what she thought it was. In fact, her paternal first cousin Pete (who also tested and was Irish) wasn't genetically related to her at all. Her world was turned upside down, and she feared she had uncovered an affair or some other family scandal.

After months of research, she eventually found a close cousin match to Pete. The match, named Jessica, was just as dumbfounded as Alice was upon the results of her discovery—Jessica expected to be European Jewish, but turned out to be Irish. What's more, they learned that Jessica's grandfather had the same birthday as Alice's father. Then the truth emerged: Alice's Irish father was switched at birth with a Jewish baby, Jessica's grandfather. The Irish baby went home with a Jewish family, and that Jewish baby went home with Alice's. Ethnicity gave Alice the insight she needed to learn more about her true genetic history.

Though your case may not be quite as interesting as Alice's switched-at-birth story, AncestryDNA's ethnicity estimates can provide you with enlightening research leads that tell you where you may have come from.

How do they work? Ethnicity estimates, to put it simply, are crowd-sourced. People from specific regions (Poland, Asia, Scandinavia, the British Isles, etc.) don't necessarily have DNA that is totally unique from other parts of the world. AncestryDNA uses data from test-takers with deep, documented, historical roots in one of twenty-six geographic regions (image **A**), tested against more than 150 regions **<support.ancestry.com/s/article/DNA-Regions>** that are based on ancestral migrations within these major regions. These populations (called reference groups or reference panels) are then compared to new test-takers, and your ethnicity estimate is a report with how closely you match each region's reference panel.

Note that the ethnicity information provided by AncestryDNA is just an estimate. It should give you a good indication of who you are and where you've come from, but don't hold it up as gospel. Some estimates may be so small that they're dubious at best, and your results may vary from testing service to testing service (on one service, for example, I am

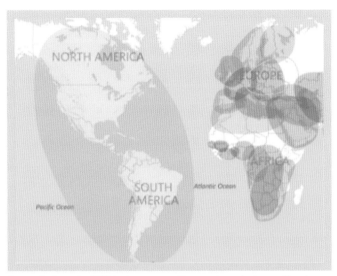

Image A. AncestryDNA tests your genetic sample against more than two-dozen reference populations in more than 150 regions.

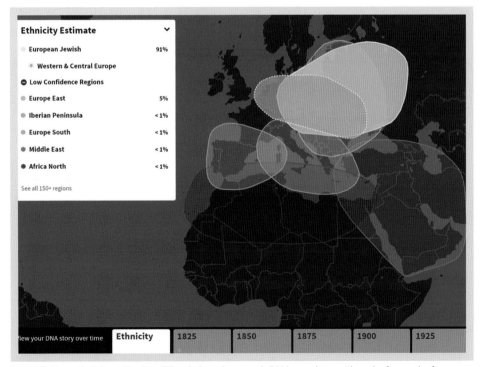

Image B. Your ethnicity estimate will break down how much DNA you share with each of several reference populations. Mine are overwhelmingly European Jewish, but I have a small percentage match with Europe East and even smaller amounts with the Iberian Peninsula, Italy/Greece, and the Middle East.

0.8 percent Eskimo/Inuit, a representation I haven't found in my results from any other services). In addition, your results may even seem to contradict each other. Siblings may have different ethnicity percentages based on what they've inherited from each parent, and you might find head-scratching relationships between your estimates and those of your parents. On AncestryDNA, I am 91 percent European Jewish, but my mother is 95 percent European Jewish and my father is 94 percent European Jewish. With so much Jewish DNA between my parents, how could my percentage of Jewish DNA be *lower* than theirs?

Again, ethnicity estimates are only an approximation. While you should take them with more than a grain of salt, the data is more for informational and recreational purposes than it is for hard genetic research.

Let's take a look at an example. As seen in image **B**, my ancestry is heavily rooted in European Jewish regions—specifically Poland, Belarus, Ukraine, Russia, Hungary, and (as AncestryDNA says) "Israel." This more or less lines up with what I already know about my heritage. My maternal side is (as far as I know) all Polish and Russian, and my paternal side is Hungarian and Polish. But I know my father's side is also Lithuanian, which Ancestry categorizes as "Europe East"; I only have 5 percent of that DNA. Again, entire

JUDAISM: RELIGION OR ETHNICITY?

Judaism is a religion. How can it be an ethnicity too? After all, there isn't a Muslim or Christian ethnicity. What makes those with Jewish ethnicity so special?

This is a question I often hear in DNA groups, so let's set the record straight. People of European Jewish ethnicity, also known as Ashkenazi Jewish, stem from a group of people who observed certain rules and regulations from a certain location. Jews of the past were confined to certain areas, with their ability to marry and even keep records strictly controlled by governments. (Even today, it's harder to establish a genealogical paper trail for those who are Ashkenazi Jewish because many Jewish records were destroyed during wars and in hate crimes.) These individuals, isolated from the rest of their communities (often by law), married within the tribe for thousands of years in a practice called endogamy.

This isolation and adversity created within them a unique community that developed a culture and identity all its own, one that transcended political boundaries and affiliations. As a result, Jewish identity became about more than religious traditions. Nowadays, a person may be Jewish by descent, but not necessarily by religion. In fact, AncestryDNA and other testing companies have identified reference panels for those with Jewish ancestry.

estimate references my ancestry back into the thousands of years, so it might not exactly
line up with what I know about the last few generations of my ancestry.

Like Alice in the *Washington Post* article, you might discover your ethnicity doesn't
line up with what you expected—though usually for less dramatic reasons. First, you may
not have inherited traceable amounts of DNA from an ancestor of a particular heritage.
DNA "dilutes" from generation to generation. Distinctive DNA from any particular ancestor or heritage gets washed out as it gets mixed throughout the generations, as you receive
about 50 percent of your DNA from each parent, about 25 percent from each grandparent,
about 12.5 percent from each great-grandparent (and so on). Through the generations, the
DNA of older generations is much less visible, until eventually, the ethnicity and DNA of
older generations are not represented in your DNA at all. In addition, you may also have
received different amounts of DNA from each ancestor, and that may skew your results—
even when comparing to a full sibling.

Lack of a particular heritage's DNA in your results may also suggest that you're not
related to that ethnic group by blood, but rather by adoption or marriage. For example,
some individuals were admitted to Native American tribes despite not having Native
American ancestry. Or, perhaps, your family's stories of a particular "heritage" (say,
Native American) resulted from living alongside those ethnic groups, rather than being
related to them.

In addition, some ethnic groups may be underrepresented. AncestryDNA's test-takers
are overwhelmingly of European and white descent, so ethnic groups from other parts of
the world may not be well-defined enough to yield accurate results. Some test-takers may
also have multiethnic backgrounds that make it difficult to parse out which DNA came
from which ethnic group.

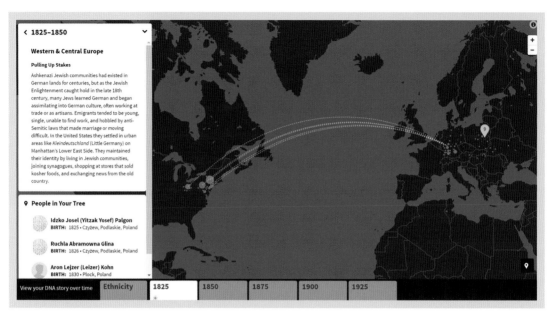

Image C. Migration groups allow you to trace your ancestors' immigration patterns throughout the centuries.

Ancestry also offers a feature called Migrations (formerly Genetic Communities) that goes into more detail about where your most recent ancestors may have come from based on specific time frames. Included as part of your ethnicity estimate, these Migrations groups identify specific historical groups of people who shared DNA and emigrated across countries together. These can provide an even more precise look at your ethnic origins. As you can see in image **C**, my family has relocated to various northeast cities in the United States, and emigrated from various countries in Western and Central Europe around 1825. Like my more general ethnicity estimate, this seems to fit well with what I already know about my ancestry.

DNA Matches

Now let's turn our attention to the most useful facet of your genetic results: DNA matches. As reinforced throughout the book, DNA does not lie—but you may run into some trouble if you don't have all the facts (or, as we've discussed earlier in this book, if you come from an endogamous culture). This section will help you troubleshoot problems with identifying and connecting with the DNA matches AncestryDNA provides.

Image D. These are my AncestryDNA match results, organized by predicted relationship. I've noted my actual, genealogical relationship with each in red. Note that AncestryDNA did a pretty decent job of estimating my relationships to each of these users.

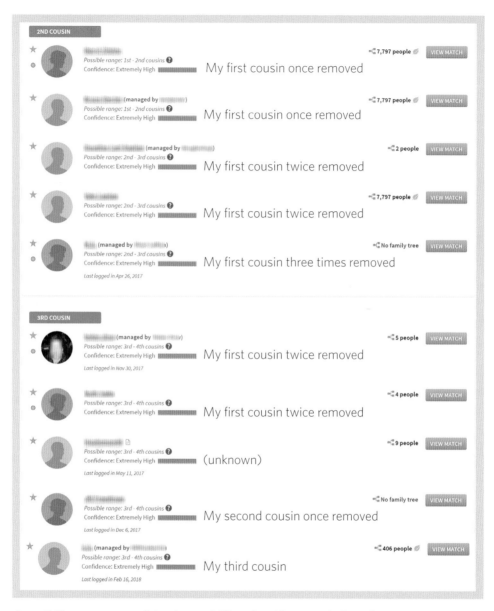

Image E. These are my more-distant AncestryDNA results, with my actual relationships to each user noted in red. As you can see, the estimated relationships here aren't as accurate as they were for more closely related matches.

AncestryDNA's uses an algorithm called Timber when calculating DNA matches. The formula ignores sections of DNA shared by thousands of people over several generations, focusing instead on the more relevant and recent matches that indicate a genetic relationship within the last few generations. By carefully selecting which genetic relationships to analyze, AncestryDNA's Timber analysis provides some of the most accurate results in the industry. This makes AncestryDNA a favorite among test-takers because it is mindful of the matches it shows.

In addition to identifying other test-takers as DNA matches, Timber also estimates how those individuals fit into your family tree. Based on the amount of shared DNA between you and a match, Timber will guess what relationship you and another user are likely to be. And because Timber is so accurate, your relationship estimates are generally reliable, too, especially at the first- or second-cousin level.

Having said that, these relationship estimates can gloss over facts or misrepresent relationships. While a parent/child estimate will likely be accurate, a first-cousin match could be a number of relationships: a first cousin, uncle, aunt, or even half sibling. You'll need to contact the other test-taker to learn for sure, and (hopefully) he'll be willing to work with you if you have such a large amount of shared DNA. Before you make any attempts at connecting, however, follow the guidelines in chapter 9's outreach section.

Let's look at an example to show how well AncestryDNA can predict established relationships. Images **D** and **E** shows my DNA match results for both close and more distant relatives. (Names have been blurred for privacy reasons.) The matches are organized by the predicted level of closeness, and the red text indicates how I'm actually related to each person.

AncestryDNA has accurately placed my parents and son as parent/child matches, and it suggested that my grandparents and uncles were first cousins. As in the chart in chapter 4 shows, first-cousin matches share the equivalent DNA that a great-aunt or great-uncle shares. You will see that they are in a family tree with 7,780 people because their tests are all associated with my account and linked to my family tree.

RESEARCH TIP

Temper Your Expectations

If you're a member of an endogamous culture, you may find you have thousands of DNA matches. However, many of these "matches" aren't as close as you think. Testing companies have no choice but to compare you with other members of your "tribe," all of whom share more DNA with you than is typical. This will inflate the number of your relationships, especially of your close relationships. Endogamy also invalidates the effectiveness of the Shared Matches feature on AncestryDNA, because you'll likely get maternal and paternal results on any search.

Now let's look at slightly more distant matches and see how they fare up. AncestryDNA, for the most part, did well. I personally know all of my second- and third-cousin matches, and I was thus able to verify those relationships. I couldn't place one "third cousin" match, however. As predicted, endogamy was at play here. This match was related to me on my mother's side and on my father's side, increasing the amount of DNA we share and essentially tricking the system into thinking we're closer than we are. We're related, but at a much more distant level than third cousins. Interestingly enough, that third-cousin match is an even closer match to my son than to me, which means she's also related to him through my husband! Going further down the match list brings me to murky territory. I just don't know who many of them are.

Despite being less accurate than closer relationships, the relationships AncestryDNA flags as more distant can still be critical to research. For example, my cousin Kim attempted to discover her ancestry and found that a relative, Doug, was labeled in the "fourth- to sixth-cousin" range by AncestryDNA when (in actuality) he was her second cousin once removed. That match could have easily fallen off her radar. My mother and her two brothers, on the other hand, are Kim's third cousins, and all of them share more DNA with each other than they do with Doug, a closer genetic cousin. (Thanks again, endogamy!)

Some of this is due to the amount of shared centimorgans (cM) between these matches. A third-cousin match in Kim and my mother's case shared 50 cM across six segments, which doesn't show much promise (especially in endogamous cultures). However, we happened to know Kim was only one-fourth Jewish and so knew which ancestor to use to trace her DNA. On the contrary, Doug and Kim (a second cousin, once removed match) shared 30 cM across five segments. But this is right where the chart puts them, with second cousins twice removed sharing between 30 and 215 cM and third cousins sharing between 0 and 109 cM.

Now that we're talking numbers, let's discuss how to view this kind of data from AncestryDNA. We've already looked at my list of matches—each of these matches are clickable, and you can glean some helpful information from each match's page (image **F**). Namely, you can view the amount of shared cM and a family tree, if available (and public).

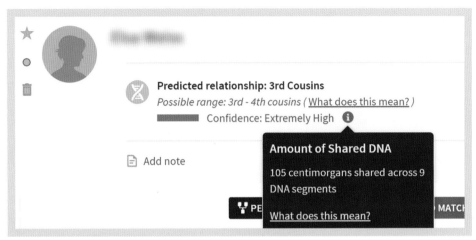

Image F. On an individual match's page, you can view AncestryDNA's predicted relationship based on the number of shared cM by clicking on that information bubble (i). In reality, this relative is my first cousin three times removed.

Tip: If you don't see a tree linked to the user on this page, you can click their avatar above to access their public profile, and there you might find additional trees created by the user.

On each individual match's page, you'll be able to view your predicted relationship to that person, plus AncestryDNA's confidence rating. Hover over the (i) to view how many cM you share with the match. You can also add text notes, which may be helpful as you go about your research and discover more about each match. For example, since Ancestry.com's messaging system doesn't work very efficiently (outbound messages are hard to find), you may want to mark that you contacted so and so on a certain date. You might want to add other information, like a last name or how you've discovered the match is related to you.

Scroll down, and you'll be able to view a number of helpful features if your match has uploaded a tree to the site. Under Pedigree and Surnames, you can view how you and your

RESEARCH TIP

Save Before You Send

Before you contact a match (particularly a close match), grab a screenshot of the match's information and (if possible) create a mirror tree (see chapter 9). I've heard hundreds of stories of people making initial contact with a potential relative, only to be blocked by the other user or see the other person's Ancestry account shut down. Some people test only to learn their ancestry and don't want to be contacted by potential relatives, especially birth parents or unknown or long-lost children.

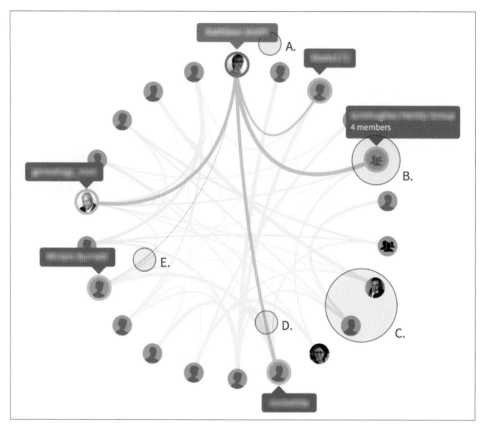

Image G. With DNA Circles, you can view a web of matches to see how they all relate to each other. Bolder lines indicate a stronger relationship.

match's tree line up, if at all. If they exist, the site also provides a list of shared surnames, helping you pinpoint which ancestors you share with the match.

The second tab is for Shared Matches, which generates a list of DNA matches you and your match have in common. This can be extremely helpful to identify shared ancestry, and can be a powerful tool used in conjunction with the Pedigree and Surnames tab. As with predicted relationships, the Shared Matches tab will be tricky if you come from an endogamous culture.

Note: The family relationships under the Shared Matches tab (e.g., "Close Family," "2nd cousin") are for how that match is related to the listed test-takers, not how you are related to those test-takers. For example, you may be a second cousin to Match A, but the person you're comparing (Match B) to is a fifth through eighth cousin to Match A. Under the Shared Matches tab, Match A will be listed as a fifth through eighth cousin, since it reflects Match B's relationship to Match A, not your relationship to Match A.

Finally, the Map and Locations features will plot the locales associated with you and your match's family trees on an interactive map. Use this if you have ancestry from a diverse set of places, or if you want to figure out where in the world your ancestors or matches' ancestors came from. And if your matches have provided detailed information (such as city or country of origin) that lines up with your own research, you may have struck gold.

All your DNA matches will feature a link to contact the match. Use this to request access to your match's family tree if they're family information is listed as private. Even if the other user doesn't have a family tree on Ancestry.com, he might be willing to provide you with some information about your potential connection. Again, consult the outreach section of chapter 9 before reaching out.

DNA Circles

Your DNA results page also features another useful tool: DNA Circles (image **G**), which matches your DNA to those of other test-takers who share DNA and have common ancestors in their family trees. You're placed in DNA Circles based on AncestryDNA's shared ancestor confidence score, which in turn is dependent on how you and other test-takers respond to Ancestry.com's questions about whether you're related to a particular ancestor. To summarize: The more people sharing the same ancestor, the more likely you will see them in a DNA circle.

You can use DNA Circles to find potential cousin matches as well as cousins of your genetic cousins. This can be a great tool for discovering shared ancestry and expanding your list of potential family members—and, as we'll see in chapter 11, it's extremely useful in triangulating relationships.

Note that the number of DNA Circles you're a member of may change over time. AncestryDNA continually updates its algorithms and refines its data, so you may find you're no longer a member of a DNA Circle (or that you've been placed in a new one). It is common to discover a Circle one day, only to see it disappear the next. Or you could have dozens of Circles and see fewer a week later.

6

Family Tree DNA

After AncestryDNA **<dna.ancestry.com>**, the most popular DNA testing service is Family Tree DNA **<www.familytreedna.com>**, run by the Houston-based company Gene by Gene. Though smaller than AncestryDNA, Family Tree DNA still boasts a database of more than two million test-takers—and critically is the only big testing service to offer accurate mitochondrial (mtDNA) and Y-chromosomal (Y-DNA) tests. Because of this, Family Tree DNA has carved out a niche for itself as the foremost provider of haplogroup information.

Family Tree DNA is also notable for its approachability and collaborative features. As a test-taker, you can join haplogroup projects, surname projects, and more, allowing you to coordinate with like-minded genealogists and others interested in exploring their history. Then, you and others can compare data (project admins can see quite a lot, helping others connect the dots) and communicate amongst one another. And since people choose to join these projects, they're generally willing to participate and collaborate.

Family Tree DNA is particularly helpful to those testing overseas. Its low cost and international shipping give it an edge, so you can expect more foreign matches there, which is helpful if you believe you were born outside the United States. DNA is still gaining traction internationally, but your test will at least be in the database if and when your foreign relatives decide to test.

Another big plus of Family Tree DNA is that the company saves your sample so you can upgrade your test later if you'd like. For example, if you take Family Tree DNA's Family Finder test (its autosomal test), you can sign up for an mtDNA, Y-DNA, or Big Y test without having to swab again. In addition, Family Tree DNA works with the swab (rather than a tube), so it is user-friendly to those with dry mouths who just can't get a good sample on other services.

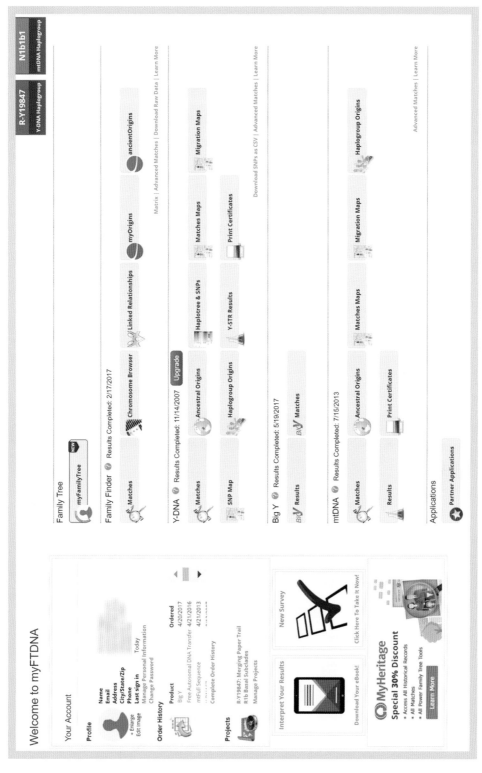

Image A. You can access the results from all your Family Tree DNA tests from your dashboard. A full dashboard (for a male who has taken all of the basic tests, plus the Big Y) will show you your mtDNA haplogroup, Y-DNA haplogroup, Family Finder matches (from your autosomal test), Y-DNA matches and results, mtDNA matches and results, and Big Y results.

Now that you know why people would want to use Family Tree DNA, let's see what we can learn with the service. We'll explore Family Finder, mtDNA, Y-DNA, and Big Y.

The Family Tree DNA Home Page

Let's take a look at the Family Tree DNA dashboard. Here, we'll examine my father's account, since he's taken almost all of Family Tree DNA's tests (image **A**).

The first option on Family Tree DNA's dashboard allows you to build out your family tree. You can either upload a GEDCOM (a type of file that stores family tree information, created by family tree-making software or websites), or you can manually create one from scratch. You'll probably want to build at least some semblance of a family tree that features known genetic relatives (if you've identified any and have established contact).

Trees can be useful because they can help you more easily identify potential relatives. If you happen to know paternal or maternal matches, you can link them to people in your tree. Family Tree DNA will let you know if other relatives are paternal or maternal matches based on overlapping DNA segments between you, your known paternal/maternal match, and your potential paternal/maternal match. It's quite a cool feature offered by Family Tree DNA.

Image B. Family Tree DNA's family tree function is a bit bare bones, but it can give you important information at a glance. This is my father's myFamilyTree showing which tests he's taken.

For security's sake, you can only see your matches' family trees. While this policy may be good for users who value their privacy, it also means you can't reach out to individuals who share your surname unless you genetically match them, potentially limiting your research options since not all relatives match each other. Note that your matches' profiles will indicate which tests they've taken—see image **B**.

The Family Finder (Autosomal DNA)

Perhaps the most useful of Family Tree DNA's services is the Family Finder (FF) tool. This service, similar to the autosomal DNA tests offered at AncestryDNA, MyHeritage DNA **<www.myheritage.com/dna>**, and 23andMe **<www.23andme.com>**, can help you discover other test-takers who are related to you.

DNA Matches

Let's dissect the FF match page (image **C**). The results page provides the gender and name for each test-taker the system matches you with, along with a match date that indicates when each match entered the Family Tree DNA system. If you get e-mails as well, you may be informed that you have "new matches," but when you sort your matches by date, you may find that "new match" is actually several weeks old.

The next column identifies the relationship range of a relative, based on similar guidelines as the charts presented in chapter 4. As we discussed in chapter 4 (and, briefly, in chapter 5), relationships are estimated based on the amount of DNA you share with another user. On Family Tree DNA's results page, you'll see the amount of shared centimorgans (cM), which is one of the most important factors in determining relationship range. The second piece of data, the longest block, is also useful, as it will tell you whether Family Tree DNA's estimated relationship is significant. The longer the block, the better your chances are of having a genealogical and genetic match. Similarly, the next column

> **RESEARCH TIP**
>
> **Reach Out on Family Tree DNA**
>
> Many (but not all) of Family Tree DNA's users are genealogists and DNA enthusiasts, and they are usually happy to collaborate and provide you with information. As a result, feel free to reach out to other users asking for help, advice, or genealogical information—you'll hopefully find they're willing to give it.

Name	Match Date	Relationship Range	Shared Centimorgans	Longest Block	X-Match	Linked Relationship	Ancestral Surnames	
	02/16/2017	Parent/Child	3,384	267	X-Match	Father		
	02/18/2017	Parent/Child	3,384	267	X-Match	Mother		
	04/26/2016	Parent/Child	3,384	267	X-Match	Son		
	04/14/2017	Half Siblings, Grandparent/ Grandchild, Uncle/ Nephew	1,907	123	X-Match	Uncle		

Image C. My FF DNA matches show lots of paternal and maternal relatives, including this selection here. As it turns out, the first three relationships in this list are my parents and my son—Family Tree DNA predicted the parent/child relationship before I even linked them to my family tree.

on FF indicates the presence of any X-chromosomal DNA shared between the matches. However small or large the amount of shared X-chromosome (X-DNA), the column has a value of "X-Match" if you and a match hold any X-DNA in common.

The final two columns depend on user-submitted information. Under Linked Relationship, you'll find how that match has been identified in your family tree (if linked, which we'll discuss; if not linked, you'll have an option to add the person to your tree). This is especially helpful when determining paternal or maternal cousins at a glance. Likewise, the column Ancestral Surnames lists the names a match has either provided manually or uploaded through a GEDCOM file.

A quick note about more distant relatives: Family Tree DNA's match algorithm is different than those of other testing companies, and it has often come under criticism. Whereas AncestryDNA and 23andMe use a threshold of 7 cM to identify close matches, Family Tree DNA's matches consider much smaller segments in order to ascertain relationship status. Smaller segments are often (but not always) less helpful when establishing a relationship, but Family Tree DNA acknowledges them as important segments. That means you'll see a lot of estimated second and third cousins who aren't actually second and third cousins.

Family Tree DNA's matching criteria has its pros and cons. I have tons of genetic matches (which, in general, is a good thing), but I also have many more distant matches that aren't worthwhile research leads.

Let's look at one of my matches to see how this bounty of matches is a blessing and a curse. When I first started on Family Tree DNA, my largest match was one of the most prominent genealogists in the United States, yet we couldn't figure out how we were related. In fact, she shows up on my list right between my first cousin, twice removed and my first cousin, three times removed. We share forty-two segments totaling 173 cM—but our largest segment is only 11 cM. That means the average segment size is 4.11 cM, which most testing services consider too low to be matches. In fact, scientists suggest that 33 percent of segments under 5 cM are false positives—but Family Tree DNA considers this a distant match. Ultimately, this prominent genealogist is too distant of a cousin for us to trace. We've spent a lot time trying to figure it out, but it will likely go nowhere. I've only been able to piece together my relationships with the closer matches, especially those whose largest segments are significant.

Despite our inability to connect our family trees, many of my distant matches and I share common surnames. This is partly because of the endogamy in my family tree, but also because it's not that uncommon to share surnames with someone you have no recent genetic connection to. In fact, you may not want to bother with common last names unless you have a unique location of origin.

For example, I have an ancestor with the surname "Gerber." Once, a DNA match who also had a Gerber ancestor e-mailed all of his matches who happened to have that surname (even the weak ones), thinking that a shared Gerber ancestor connected us all. Despite several having the same country origin, these matches were miles apart (too far away to travel by car, let alone by foot). There was no way to confirm the relationship was through a common Gerber ancestor, and it's more likely that each of us are connected through different ancestors entirely. That mass e-mail wasn't received well by some recipients; it is often a newbie mistake to assume that the shared last name is definitely the connection. Even with a recognizable surname, the relationship could be through an unknown ancestor.

Moral of the story: Don't e-mail everyone with a particular surname on a whim unless they are close enough to establish a relationship (or unless the surname is unusual). Having said that, keep your eye out for matches who share surnames and locations with you. Grab their family data, then contact them via e-mail using the blue envelope icon under the match's name (following the guidelines outlined in chapter 9). You can also leave a note to yourself to indicate any past correspondence with a match (or for any other purpose).

Additionally, you can see at a glance if a match has uploaded a family tree. A dark blue icon indicates a family tree, but (of course) you can still attempt to get genealogical data from a match who hasn't uploaded a tree by contacting them. You can also click a match name to find out a variety of information, depending on what tests a person has taken and what information he has provided: name, e-mail address, haplogroup(s), earliest known ancestors, biographical information, and ancestral surnames.

Assuming you've established a connection with another test-taker, linking relatives to your family tree is easy. Once you've entered their name in your tree, you can connect the name from Family Tree DNA's results (seen on the left) by dragging the result directly on top of the person in the tree (image **D**). Since this assumes they're the same person, select "Self" (image **E**). If it's a new relative you just discovered, you can choose one of the other options.

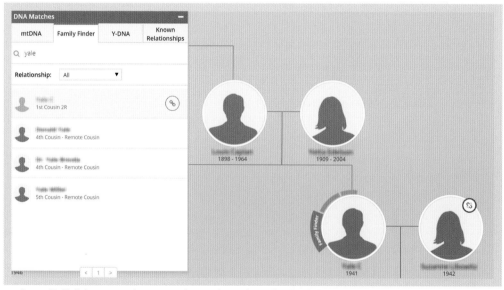

Image D. To link your matches to specific members of your family, click the name in your search results and drag it to the person in your tree. You'll then see the image above, which lets you identify your match. If your tree is already uploaded and you see the match in the search results as well as the tree, drag the match name in the search results over to the name in the tree and select Self. If it's a new relative you just discovered, you can choose one of the other options based on their relationship to the match.

Image E. When you drag a match to someone in your tree, you're likely adding them as Self, but they could be other relatives, too. Drag the person to the right name. If you haven't yet made contact with this person, you'll probably want to uncheck the option to send a notification e-mail.

The Chromosome Browser

In addition to providing matches, FF can generate another incredibly powerful analysis tool: a Chromosome Browser. The Chromosome Browser lets you compare up to five relatives to see if there's any DNA overlap, and the tool even breaks down how much DNA you share with each relative on individual chromosomes. Select the users via the checkbox near each match's avatar. Then scroll to the top of your match list and click Chromosome Browser. This option is available for free if you've taken a Family Tree DNA Family Finder test, but you'll have to pay a nominal fee to unlock this feature if you're uploading your test results from another company.

When accessing the Chromosome Browser directly, you can filter relatives by immediate relatives, close relatives, close and immediate relatives, distant relatives, speculative relatives, confirmed relatives, common surname, new matches since last sign in, name (last, first), X-matches, and all matches. The option you choose filters the list into more digestible segments (unless you choose all), and from there, you can select up to five relatives to do the same comparison.

See image **F** for the Chromosome Browser view. In this illustration, I'm comparing my DNA to that of my uncle, my mother, my mother's first cousin once removed, and my mother's paternal first cousin. Each of the four colors represents one of those four

Image F. Family Tree DNA's Chromosome Browser, which is an invaluable tool when comparing your DNA to others, is available to those who bought FF tests or paid to unlock it when uploading results from another company. Each color represents the DNA you share with a match on individual chromosomes.

relatives, and the colors that appear represent the DNA I share with each on that particular chromosome. I set the match threshold at 10 cM because I know these individuals are very close relatives to me, but Family Tree DNA allows you to set the threshold as low as 1 cM.

See how the blue line overlaps entirely with my DNA? Blue represents my mother, and I share so much DNA with her because (after all) I inherited half of my DNA from her. By comparison, the orange line (my uncle) shares less with me, and partially (but significantly enough to confirm they are full siblings) with my mother (the blue). Even though my mother and uncle are full siblings, they don't share the exact same DNA because of the randomness with which DNA is inherited. The green line is my maternal first cousin twice removed, who doesn't overlap with my DNA at all on chromosome 1 and only slightly does on chromosome 2—in an area that my mother shares, but my uncle doesn't (which, again, is attributable to the randomness of inherited DNA). My paternal first cousin once removed (in pink) overlaps with my mother and uncle in many places, but not with my maternal cousin. To put it another way: My mother's maternal and paternal cousins don't seem to have any overlapping DNA, meaning endogamy isn't at play when looking at these particular relationships. Because of this, as they're closely related to me,

I can almost definitively attribute these overlapping segments to the most recent shared ancestor on both my maternal and paternal sides.

Note that there are tiny blue links on the top of the Chromosome Browser ("Optional Views") that may prove useful in the rest of your search. You can use these to download your matches (either selected or ALL) to an Excel file, where you can begin a deep analysis of your matches to see where there is chromosomal overlap. This may prove particularly useful in finding parentage, especially if there is segment overlap by known cousins in certain areas.

The Chromosome Browser CSV (comma-separated values) results will show you your match, the chromosome the match is on, the start location of the match, the end location of the match, the number of cM shared across this entire region, and the matching single nucleotide polymorphisms (SNPs). Using this information can help you limit results to significant matches and can be read offline. Other services can read this data to give you more actionable information—see chapter 10 for more on this. This information will become critical later as you attempt DNA triangulation, in which you compare three people who have overlapping segments in the same place (see chapter 11).

The In Common With (ICW) Tool

Another feature is the In Common With (ICW) tool that appears on the top of the page next to the Chromosome Browser button. This allows you to identify additional individuals who you and a match both share DNA with. This is great for identifying additional test-takers/matches to investigate, as well as for expanding your family tree. For example, I selected my mother, then clicked on the ICW button. Doing so brought up all my maternal relatives (or at least the ones we think are). This tool can be used in conjunction with a surname search, as well as to narrow down possible reliable connections. It's important to understand, however, that the query results don't match on the same segment. But their DNA overlaps with me, my mother, and the test-takers who show up in the results.

This feature can be a godsend. I can easily isolate maternal matches here, but when using the Not ICW feature, I can find relatives who are not in common with my mother. Let's say I know my mother but want to discover who my father is. Clicking Not ICW is a good place to start.

As with many DNA topics, there's a caveat for endogamous cultures. When you go further down the list, you may find ICWs that are both maternal and paternal. My paternal third cousin once removed is coming up as a maternal match, too. That means he is related to me through both parents, but I've yet to discover how he matches my mother. Chances are, it's too distant for us to find.

Ethnicity Estimates: myOrigins and ancientOrigins

Like AncestryDNA, Family Tree DNA provides you with information about exactly where in the world your DNA comes from. The first feature, myOrigins, assigns you a profile based on twenty-four reference panels from around the world. The results tag me as 98 percent Ashkenazi Jewish, with trace regions in East Europe and North Africa (image **G**). You can also view your results in map form by clicking the red text that says "View myOrigins Map" (image **H**).

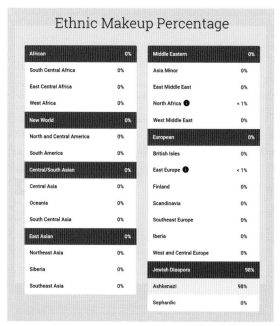

Ethnic Makeup Percentage

African	0%		**Middle Eastern**	0%
South Central Africa	0%		Asia Minor	0%
East Central Africa	0%		East Middle East	0%
West Africa	0%		North Africa ⓘ	< 1%
New World	0%		West Middle East	0%
North and Central America	0%		**European**	0%
South America	0%		British Isles	0%
Central/South Asian	0%		East Europe ⓘ	< 1%
Central Asia	0%		Finland	0%
Oceania	0%		Scandinavia	0%
South Central Asia	0%		Southeast Europe	0%
East Asian	0%		Iberia	0%
Northeast Asia	0%		West and Central Europe	0%
Siberia	0%		**Jewish Diaspora**	98%
Southeast Asia	0%		Ashkenazi	98%
			Sephardic	0%

Image G. When I click Show All on the myOrigins page, I get a breakdown of my (boring) ethnicity. Your results will probably be a little more diverse.

Image H. myOrigins will also map out where your DNA comes from. I'm from right where I expect to be, and other matches of mine are from there, too. You can also view matches who come from the same regions—I only recognize one name on the Shared Origins list, and that's my father.

In addition to these more-conventional myOrigins results, Family Tree DNA offers a tool called ancientOrigins. What's the difference? ancientOrigins tells you what types of ancient peoples you come from: Hunter-Gatherers, Metal Age Invaders, Early Farmers, etc. The map itself tells you even more by tying each of these trades to a specific location (image **I**). While none of this truly helps with finding your immediate family, you might find these insights about your deep ancestry interesting.

The Matrix

Before we move on from the FF test, we need to discuss the Matrix. The Matrix is a tool that lets you look at several matches on a grid, showing whether or not they're matches with each other. Like the ICW tool, this can be useful in identifying potential relatives to explore and understanding the connections between your various genetic relatives.

Image **J** shows three of my paternal relatives, plus one unknown match who I'm trying to identify. What's interesting is that the unknown match (fourth column and fourth row) seems to match with my paternal grandfather, but not my grandfather's paternal first cousin or that man's son. This means either the match is endogamous or she is related to him but through his maternal side. I'd have to analyze further to find out.

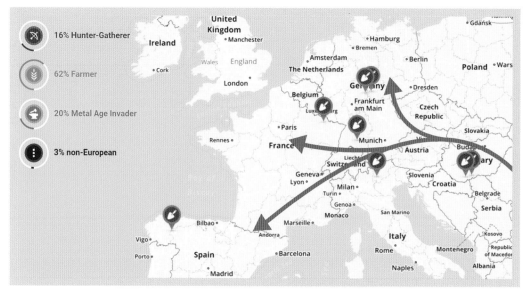

16% Hunter-Gatherer

62% Farmer

20% Metal Age Invader

3% non-European

Image I. While ancientOrigins probably won't help you find recent ancestry, you can still view what your ancient family's trade may have been doing millennia ago. The map shows me ancient migration patterns for the Hunter-Gatherer group.

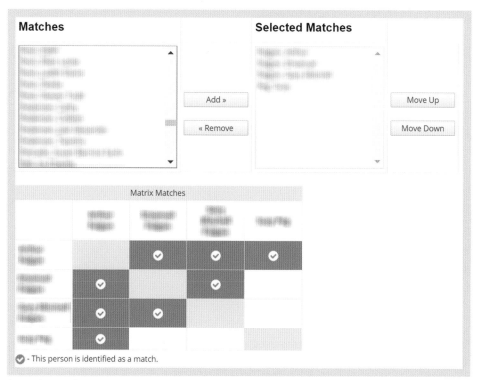

Image J. I can choose up to ten of my DNA matches and compare them all using the Matrix tool. This will tell you which of your DNA matches match with each other.

mtDNA

While the FF is arguably the most useful of Family Tree DNA's tests, let's now look at the mtDNA test, which examines the deep ancestry of your maternal line.

While it offers only one kind of autosomal DNA test, Family Tree DNA offers three different testing levels for mtDNA: mtDNA, mtDNAPlus, and mtFullSequence. Each test samples a different combination of DNA regions (HVR1, HVR2, and the full Coding Region), with the basic mtDNA test only analyzing one specific region. The biology behind these three regions is beyond the scope of this book—all you need to know is that these different testing levels will give you different match information. The more DNA you test, the more you're able to learn about your genetic makeup and how you match up with other mtDNA test-takers.

Testing level	Matching level	Generations to common ancestor	
		50% confidence interval	95% confidence interval
mtDNA	HVR1	52 (about 1,300 years)	n/a
mtDNAPlus	HVR1 & HVR2	28 (about 700 years)	n/a
mtFullSequence	HVR1, HVR2, & Coding Region	5 (about 125 years)	22 (about 550 years)

When you view your mtDNA results on Family Tree DNA, you'll see your match information. Similar to Family Finder, your results include:

- a name you can click, which displays a small popup with identifying information and other information that individual has shared
- an icon that you can use to e-mail the match
- a note icon for you to leave notes about your match.
- a tree icon, which indicates that your match has uploaded a family tree

Then, you'll see which tests you've taken: FMS refers to the mtDNAFullSequence, FF refers to Family Finder, and HVR2 refers to the mtDNAPlus test.

You'll also see the genetic distance between you and your match. You may want to review you and your match's Family Finder results to see if you're closely autosomally connected if you have a low genetic distance between you. If not, the match is likely distant and harder to determine.

The match list also shows other helpful information, depending on what the test-taker provided: the earliest ancestor and his approximate birth date, the test-taker's haplogroup, and a match date (the time that the match was added to the system).

Image K. Family Tree DNA maps out where other members of your haplogroup currently live. From what I can see, none of these folks are exact matches, but instead have one or two mutations and thus are not recent genetic cousins.

You can sort your results by new matches, ancestry, haplogroup, and match date. You can also filter by which kind of test was taken (e.g., HVR1 or HVR1 & HVR2). When you do, the haplogroup listings may include individuals who are in a broader haplogroup. For example, my haplogroup is U4a3a, but doing this shows me U4a3 and U4 matches as well.

One final thing about this page: You can also search by surname or matches that are new since a certain date. Surnames may prove to be somewhat difficult as maiden names typically change throughout the generations.

mtDNA also gives you a breakdown of your mtDNA haplogroup, which provides a clue to your maternal ancestral origin. You can view where you and other members of your haplogroup currently live (image **K**). This map shows diversity in modern-day haplogroups. For example, I have DNA matches from my haplogroup in both Afghanistan and Germany, as well as the United States. It might not be totally accurate (Family Tree DNA creates this data from user-reported information, even though we know U4 has Indo-European origins and is prevalent in Finland and Russia), but is still fun to look at.

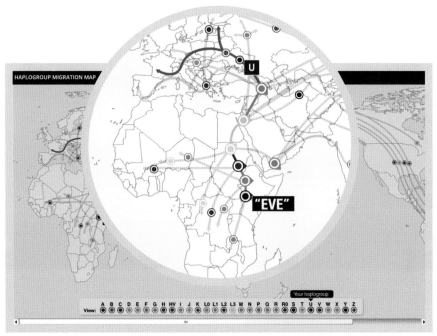

Image L. Family Tree DNA provides you with a map that tracks the migration of your haplogroup, from Mitochondrial Eve to present. Another view can tell you how prevalent your haplogroup is in the modern populations of various regions.

Family Tree DNA also provides migration maps that show how far your haplogroup has spread over the centuries. You can see the paths of migration in image **L**, with the labeled brown line indicating how the U haplogroup moved out of Africa, through the Middle East, and into Europe. You can learn more about each haplogroup's by clicking the Frequency Map tab.

You can discover even more about your mtDNA haplogroup under the Results tab. Here, you can see the specific mutations that characterize you within your haplogroup (image **M**). This page breaks down which mutations you have on each of the coding regions you tested, as compared against the RSRS (Reconstructed Sapiens Reference Sequence, a constructed human genome). A lot of this will likely go over your head, but someone in your haplogroup may contact you if they want to study your haplogroup further. (That's how I was persuaded to ask my father to take a Big Y test!)

Mostly for fun, Family Tree DNA offers certificates about your haplogroup and origins. Print them out and proudly display them in your home or office! It's a cute feature and allows you to display your genetics with pride.

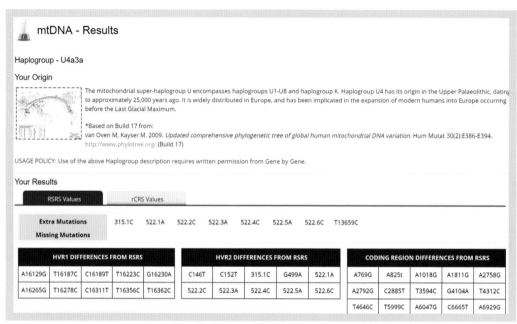

Image M. You'll view your mtDNA results as a list of differences from the RSRS (Reconstructed Sapiens Reference Sequence), which is a constructed human genome used when comparing mtDNA.

For more on exploring your mtDNA results, check out the International Society of Genetic Genealogy (ISOGG) Wiki **<isogg.org/wiki/MtDNA_tools>**, which can connect you to other websites and applications when analyzing your mtDNA data.

Y-DNA

Family Tree DNA offers one last category of DNA tests: Y-DNA. Like mtDNA, your first option is to look at your matches. As we've discussed previously, Y-DNA allows you to test down your paternal line, as Y-DNA is passed from father to son. Like mtDNA, your Y-DNA can tell you about only one line of your ancestry (in this case, your father's father's father's father, and so on). As a reminder, only men can take a Y-DNA test, so women will need to ask a father, grandfather, uncle, male cousin on the same paternal line, or brother to test. You can learn more about Family Tree DNA's Y-DNA test on its website **<www.familytreedna.com/why-ftdna.aspx>** or on ISOGG Wiki **<isogg.org/wiki/Y-DNA_tools>**.

As with mtDNA tests, Family Tree DNA offers various levels of testing, each testing a different number of regions. The more regions you test, the more comprehensive your results. At the time of this book's writing, Family Tree DNA offers three Y-DNA tests: Y-37, Y-67, and Y-111. (The company also previously offered Y-12 and Y-25 tests.)

Image **N** shows the results of my father's Y-DNA67 test, which (as the name implies) examines sixty-seven regions of Y-DNA. Note that he has matches with a genetic distance of 0, which would indicate close genetic relatives. In fact, he has relatively few matches at all—none of them share his surname, though their shared haplogroup (R-M269) is common to European Jewish ancestry.

As with mtDNA results, you'll see a list of matches providing a variety of information: genetic distance, name of match, and icons to e-mail the match, add notes, and see a family tree (if available). In addition to these, the Tip button tells you how likely a match has shared a common ancestor within the past several generations (four, eight, twelve, sixteen, twenty, and twenty-four by default, but this can be changed and recalculated if you want to fine tune your results on a generational or every-other-generation level). You can also see what other tests the match has taken, including the type of Y-DNA test (e.g., Y-67) and info about other kinds of test (e.g., the Family Finder). And like with mtDNA, you can see the earliest known ancestor (along with approximate birth years) if the user has provided this detail. The Y-DNA haplogroup column follows, and (like mtDNA) this encompasses your specific haplogroup plus the parent haplogroups from which your haplotype has emerged. Your results may also list a terminal SNP of the latest subclade known to be researched. Finally, you can see the date the match was established.

You may be able to narrow down results to basal subclades and groups that have merging paper trails (if available), as well as search fewer (or more) markers. Search options allow you to look for surname and recent matches from a certain date.

As with mtDNA tests, Y-DNA tests also provide you with a breakdown of your ancestral origins. You'll see a chart of your Y-DNA matches, where they come from, and the

Genetic Distance ↑	Name	Earliest Known Ancestor	Y-DNA Haplogroup	Terminal SNP	Match Date
2	Y-DNA67 FF	Mr.	R-M269		4/8/2015
2	Y-DNA111 FF	b. 1756 and d. 1825	R-M269		5/9/2012
3	Y-DNA67 FF	circa 1797 London, d 1865Hobart,	R-M269		11/1/2017
3	Y-DNA67 FF		R-M269		6/26/2017
4	Y-DNA67 FF BigY	b abt 1762	R-A13358	A13358	12/6/2015
4	Y-DNA67 FF		R-M269		11/27/2014
4	Y-DNA67 FF	b. 1760 and d. 1834	R-M269		9/10/2013
4	Y-DNA111 FF BigY	1925-2014, Hersh Pollak Apter 19	R-A13358	A13358	7/11/2012

Image N. My father's Y-DNA results are sorted by genetic distance. Based on which Y-DNA test you take, you'll be able to review Y-DNA matches at different marker thresholds. Since my dad took a Y-67 test, he can also view his results at a variety of marker thresholds: 12, 25, 37, and 67.

THE BIG Y

We will go into Big Y briefly, only because those curious may want to see what it looks like. Essentially, the Big Y test examines many more testing regions than do the other tests—about twelve million, in fact. This more detailed analysis won't necessarily help you find birth parents, but it can provide interesting insight for those interested in pursuing Y-DNA lines. In order to take the Big Y, you need to have taken one of Family Tree DNA's Y-STR tests (the number of markers you tested is irrelevant).

Once you receive your results, you'll see a list of your Big Y matches, which may be helpful in some cases but is usually more of a gamble. For example, none of the last names in my father's list of results match his, nor does he have any exact Big Y matches.

You can download your Big Y data in either of two formats. The VCF file is informational, showing all the variants tested, and the BAM file shares more in-depth data: raw sequencing read alignments. The BAM file can be used on a site like YFull **<www.yfull.com>**, which unlocks even more information about your Y chromosome data and is constantly being studied and reviewed by Big Y scientists and aficionados.

total number of test-takers who live in that specific country. You can then view a Y-DNA haplotree, where you can view how many markers you still have left to test and order additional analyses.

You can also view a Matches Map (image **O**), where you can zoom in on locations and click on the markers to learn about your matches and where they're from. Note that this tool examines your matches using just twelve markers for comparison, so you may see more matches than you expect. And, just as with mtDNA results, you can view a migration map of your Y-DNA haplogroup's travels over the centuries.

One analysis tool unique to the Y-DNA test is the Y-DNA SNP map (image **P**). This shows an extensive list of subgroups within your haplogroup, each with genetic variants at specific SNPs. You can view this data either as a dropdown of haplogroups (alphabetically, as seen in the image) or on a map of up to six SNPs. Note this tool takes quite awhile to load, because there are a lot of SNPs within Y-DNA that it is looking at in each haplogroup.

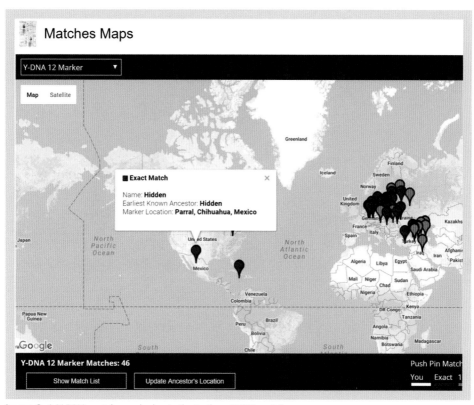

Image O. A Y-DNA match map looks similar to the mtDNA match map.

Let's take a closer look at this breakdown. The table (not pictured) shows my father's Y-DNA Haplogroup Origins. Here, I can see the specific subgroups of my father's Y-DNA matches, plus where they live. Germany seems to be the prevailing origin of these test-takers—though these locations are self-reported and could be inaccurate.

For even more detailed information, you can view your Y-STR values (image **Q**). My father tested 67 markers, so he gets results for all 67 SNPs. These are useful when comparing other test-takers' results to yours, as these can tell you which specific STR values you share with your matches. Learn more about individual markers at **<en.wikipedia.org/wiki/List_of_Y-STR_markers>**.

The last offering Family Tree DNA provides to Y-DNA test-takers is, again, the certificates. Men are issued a Y-STR certificate, a Y-SNP certificate, an "understanding your results" document, and a Y-DNA migration map. Hang them proudly to show you know where you came from.

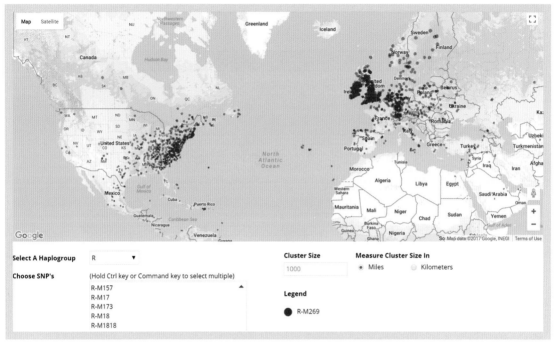

Image P. I can view on a map where members of my Y-DNA haplogroup R-M269 live.

Y-DNA - Standard Y-STR Values

PANEL 1 (1-12) ⓘ

Marker	DYS393	DYS390	DYS19 **	DYS391	DYS385	DYS426	DYS388	DYS439	DYS389I	DYS392	DYS389II ***
Value	12	24	14	10	11-15	12	12	12	13	14	28

PANEL 2 (13-25) ⓘ

Marker	DYS458	DYS459	DYS455	DYS454	DYS447	DYS437	DYS448	DYS449	DYS464
Value	15	9-9	11	11	25	14	19	29	15-15-16-17

PANEL 3 (26-37) ⓘ

Marker	DYS460	Y-GATA-H4	YCAII	DYS456	DYS607	DYS576	DYS570	CDY	DYS442	DYS438
Value	11	12	19-23	16	18	19	17	34-39	12	12

Image Q. Each marker that was tested is given a value, which you can view in detail.

Other Family Tree DNA Tools

Now that we've covered each individual kind of test offered by Family Tree DNA, let's examine some of the company's miscellaneous features.

Family Finder, mtDNA, and Y-DNA test-takers all have access to the Advanced Matches option (image **R**), which is identical under each section. It will let you compare all types of results to each other (you'll need to choose a few to compare) and you'll see where you may fit among these matches.

Family Tree DNA also has groups containing people who match each other on a surname, Y-DNA level, mtDNA level, and geographic level. In any of these groups, you can discuss and learn more with other group members, and group administrators can see your results and give you more insight into who you are in comparison to others in your group. Don't be surprised if someone reaches out to you once your results are posted and asks you to join. With the information they glean, it's possible to research further using DNA.

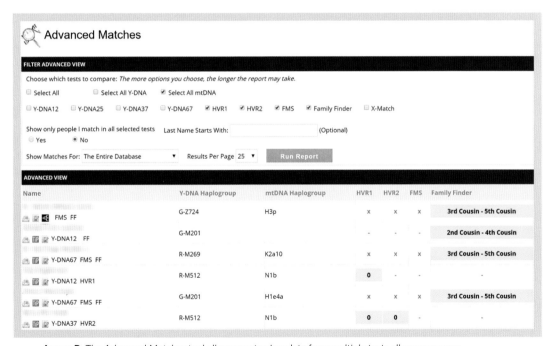

Image R. The Advanced Matches tool allows you to view data from multiple tests all on one screen.

7

23andMe

Now we move on to 23andMe **<www.23andme.com>**, a health and genetics company based in Mountain View, California. While 23andMe doesn't offer multiple kinds of DNA tests (like Family Tree DNA **<www.familytreedna.com>**) or boast a huge family tree service (like AncestryDNA **<dna.ancestry.com>** or MyHeritage DNA **<www.myheritage.com/dna>**), the company has its own unique selling point: DNA tests that provide autosomal DNA results, haplogroups, *and* health information. 23andMe's test analyzes DNA segments to see if they carry specific variants associated with genetically linked health issues. Given these health indicators, there's definitely some benefit in analyzing your DNA with 23andMe. Many adoptees want to know what their health history truly is, and the 23andMe test can shed some light on that. See the 23andMe Health and Wellness sidebar, as well as the Appearance Predictors sidebar, for more on this aspect of the 23andMe test.

23andMe's forte is ultimately the health information, leaving one with a feeling that 23andMe considers finding birth family to be an afterthought. That said, I would never advise you not to test there. Test everywhere you can if your budget allows!

DNA Matches

If you've taken only the ancestry test on 23andMe, you'll find your DNA matches under the Tools menu, where it says DNA Relatives. You'll likely spend most of your time working with this feature. This tool pairs you with other users in the 23andMe database and (like other services) estimates how closely related you are. You will likely have fewer DNA matches than the two other major services (I, for example, only have 1,247 genetic

Image A. You can sort your DNA matches on 23andMe in a variety of ways. Here, they're sorted by the calculated strength of relationship.

relatives) because 23andMe caps the matches at two thousand. Your number of matches may also fluctuate as 23andMe's algorithm changes or as users opt out of DNA Relatives

Your results page (image **A**) allows you to sort matches by one of several factors: estimated strength of relationship (close family to second cousins, second to third cousins, etc.), percent related (the percentage shared between you and your match), segments shared, and newest relatives added to the database. You can also filter by the matches you've selected as favorites.

23andMe also allows you to share more-detailed genealogical information with specific users. If someone has openly shared their entire DNA, any match can access information including shared DNA segments, surnames, and locations. Other users require you to manually request access to their DNA and other shared information before they opt in, adding another level of complication. On your results page, you can also sort by the levels of this sharing you have with other users: people who are sharing (matches who share their genealogy with you), people who are pending (matches you requested to share with you but who haven't responded yet), people who are openly sharing (those who share with the public and don't need to be asked), and people who aren't sharing at all.

23andME'S HEALTH AND WELLNESS INFORMATION

Since this book is not really about health information, we aren't going to go into too much detail about this aspect of the test, but we'll briefly overview some of the conditions the company tests for (should you opt in to health information). 23andMe looks for variants associated with genetic conditions, such as celiac disease, late-onset Alzheimer's disease, Parkinson's disease, and age-related macular degeneration. If 23andMe detects a variant, it will give you more information about the additional risk factors associated with that disease.

23andMe also indicates whether you are a carrier for a number of diseases, including cystic fibrosis, Salla disease, and sickle-cell anemia. You'll also be told if the variants they test are not detected—with the disclaimer that the analysis doesn't test all variants for each particular disease, so you may still be a carrier. 23andMe will also e-mail you with new carrier statuses as they discover them in the lab.

In addition to assessing what genetic diseases you may be susceptible to, 23andMe also analyzes other health and wellness factors associated with your genetics. For example, your results will indicate how likely you are to be lactose intolerant, how you're likely to respond to saturated fats, and how much caffeine you're likely to consume.

23andMe has a dedicated page for all of these genetic conditions, disease variants, wellness details, and (as we'll discuss later) physical traits. You can learn about each by clicking on them, and 23andMe will give you actionable information that you can use to address what it has found about you.

The health results are beneficial for those who are curious, but (as you may recall in chapter 3), don't take the results at face value. You may have the same variant as other people (some of whom have the associated diseases and conditions), but genetics are just one of many factors contributing to these conditions. Your lifestyle choices, the environment in which you live, and luck are also key in developing many of these conditions. Genes are just one part of the equation. Use these results as a guide for investigating your family's medical history, not as a diagnosis.

It's also worth noting that 23andMe's services have come under scrutiny for not adequately providing users with information about the test's accuracy, with the Food and Drug Administration (FDA) temporarily forcing 23andMe to cease providing health information. After review, 23andMe has addressed those concerns and now has the FDA's approval, though the company's users should still look at their health results with a wary eye based on other factors. That being said, 23andMe can provide information on three genetic variants found on the BRCA1 and BRCA2 genes, which are associated with the risk of breast and ovarian cancer in women, and breast and prostate cancer in men.

Let's look at my matches in detail to see what we can learn. I don't find 23andMe's default Sort by Strength of Relationship option to be all that useful, as 23andMe is guesstimating that many of these relatives are much closer than they are due to endogamy. I've only been able to identify one of my matches (the first one who I starred and manually identified as my first cousin once removed). The rest? Who knows—very likely none of them are second or third cousins, as I've communicated with quite a few and we never find any common ancestor. In fact, one of my matches shares a huge segment of his X chromosome with me, but this genetic relationship isn't particularly helpful when looking for a common ancestor since X-DNA doesn't recombine as often as autosomal DNA. Chances are, your experience with Sort by Strength of Relationship will be much better when endogamy is not in play.

My results become much more useful when I sort by percent related. I've changed the Strength of Relationship for some of my matches to their real relationships to me: my first cousin once removed, two of her children (my second cousins once removed), and other cousins who I know for a fact to be related. The guy with the big X-DNA match is still near the top of this list, too—I'll probably never determine how we're related.

From your match list, you can click the name of a match who's shared genealogical information with you to see a basic chromosome browser (which, again, is a breakdown of the specific chromosomes upon which you and your match share DNA). The view here is fairly limited. You have to go to a different part of the site (the comparison tool, also linked to your list of DNA relatives through a tiny link in the introductory sentence that says "DNA Comparison View") to access full chromosome browser information, such as start and stop points and number of shared cM. We'll review that later in this section.

Below your match's chromosome browser is some other helpful information—if your match has provided it through sharing. This includes ancestor locations you can use to compare. As with many things in genetic genealogy, this is user-reported information, so you may find your matches haven't provided data here. You can also view your match's haplogroups.

The bottom of a match's page lists surnames that these participants may be seeking more information about. They may not match with any of the surnames you're researching, but (with enough collaboration from both sides) you may be able to find the missing generations that end up connecting your surnames to theirs.

Similarly, you can also view how other DNA relatives compare to this match, helping you triangulate how you might all be related. This is one of 23andMe's best features, and it's unique among the major testing companies: AncestryDNA and Family Tree DNA just tell you how you're related to one other user, and Family Tree DNA's chromosome browser doesn't give approximate relationships of two different matches to each other. If

APPEARANCE PREDICTORS

23andMe analyzes how you're genetically predisposed to various inherited physical traits. For example, the test indicates how likely you are to have light versus dark hair and eyes, plus how many freckles you're likely to have and whether or not you have the "widow's peak" hairline. Perhaps most amusingly, your results will indicate if you can smell the asparagus metabolite in your urine. As with the health results, take these with a grain of salt, as they're only predictions based upon your genetic information. (After all, you already know what your hair and eye color are!) In fact, the trait estimates in my results were inaccurate in a number of areas: My hair is black, but was listed as "likely light." Likewise, I have a cheek dimple but I "likely [have] no dimples" according to my results. Then again, consider that these are based on reported data, where 52 percent of customers who are genetically similar to me do not have dimples. That's almost an even split!

your match has opted into sharing (whether manually or through open sharing), you can use this feature at 23andMe to potentially triangulate between relatives.

Let's look at this feature in more detail. Under the Relatives section of a match's page, you'll see a list of relatives you have in common (image **B**). For each relative, 23andMe lists how you relate to the person, plus how your match relates to the person.

Under Shared DNA, you'll see one of five options:

- Yes: You, the match you're looking at, and the listed relative have overlapping DNA segments.
- No: You, the match you're looking at, and the listed relative don't all overlap on the same DNA (though you share DNA with each of them individually).
- Share to see: The relative in question needs to share their DNA results with either you or your DNA match view this information.
- Request sent: Your results are pending, and the other user needs to respond to your request.
- Not available: The match you're looking at hasn't shared their DNA with you, so you won't be able to do any comparisons.

Each of these options is clickable, allowing you to see more detail. If you click on a name that has Yes in the sharing column, you can see what specific DNA overlaps for all three of you. Hover over the segment for more information, or scroll down to the bottom of that page for detailed segment information. If you click a name that has No in the Shared DNA column, you'll see your DNA compared to both the match you're looking at

You have **1101 relatives** in common with ▓▓▓▓▓▓

▓▓▓▓▓▓

Finding common relatives can help you piece together your family story.

Relative In Common	You	▓▓▓▓▓▓	Shared DNA
IF ▓▓▓	1st Cousin, Twice Removed 3.81%	Distant Cousin 0.40%	Share to see
RG ▓▓▓▓▓▓	2nd Cousin, Once Removed 2.86%	5th Cousin 0.15%	No
PF ▓▓▓	2nd Cousin, Once Removed 1.70%	Distant Cousin 0.09%	Sent Request
AS ▓▓▓	2nd Cousin 1.49%	Distant Cousin 0.28%	Share to see

Image B. 23andMe has a unique view that allows you to view on one screen how you and a match are both related to other users.

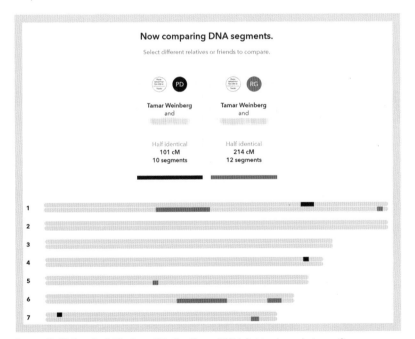

Image C. Click an individual result in the Shared DNA list to view what specific chromosome(s) you, a match, and a third relative share. Purple indicates where my DNA overlaps with one user (PD), while orange indicates where my DNA overlaps with the other user (RG). As you can see, the three of us don't all share the same segments of DNA.

Image D. 23andMe's chromosome browser allows you to compare your DNA to that of your matches on individual chromosomes. Scroll down for a detailed breakdown of start points, end points, and more.

right now and the person who is listed (image **C**). Colored DNA segments on the browser indicate what DNA you share with each of the other users, but you will not see any overlap between all three of you. The other options point you to buttons to request to share or cancel your sharing request.

In addition, 23andMe has another unique comparison view. You can see how you and multiple other users are related to each other under Tools>DNA Comparison, the DNA Comparison link on the DNA Relatives page, or directly at **<you.23andme.com/ tools/relatives/dna>**. Like Family Tree DNA's Chromosome Browser, this feature lets you choose up to five matches to compare yourself to. In this case (image **D**), I am choosing a third cousin once removed (on my paternal grandfather's side), an estimated third cousin (on my paternal grandmother's side), and a third cousin once removed (also on my paternal grandmother's side, but not related to that estimated third cousin). The display will show you which segments of chromosomes you and the other users share. Scroll down further to see start points, end points, and the number of cM shared between you and each relative. This feature will prove very useful when it comes to triangulation, which we will discuss in chapter 11.

NEANDERTHAL ANCESTRY

Neanderthals became extinct more than forty thousand years ago, but many Neanderthal's cross-bred with Homo sapiens before going extinct. As a result, you may still have some Neanderthal DNA within your cellular makeup. Unlike any other DNA test on the market at the time of this writing, 23andMe offers data that shows how much of your ancestry is traced back to Neanderthals, which comprises 4 percent of your DNA.

23andMe tests 2,872 variants and compares your Neanderthal ancestry to that of your other DNA relatives, providing a "scoreboard" of how you compare with other DNA matches. You're also given a statistic that tells you how many variants you have compared to the rest of 23andMe's test-taker population.

The section also shows you what type of Neanderthal traits you may have inherited based on Neanderthal variants tested. These traits indicate whether you have straight hair, how likely you're to sneeze after eating dark chocolate, how much back hair you may have, and details about your height that can be determined by Neanderthal variants.

This information probably won't help you find genetic relatives, but it might make for some good conversation once you do connect with your family!

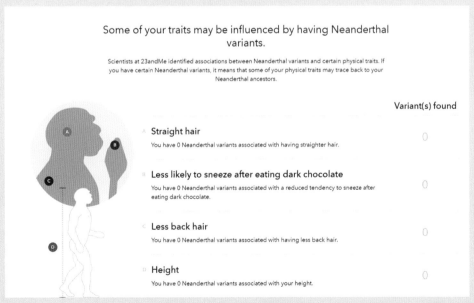

Some of your traits may be influenced by having Neanderthal variants.

Scientists at 23andMe identified associations between Neanderthal variants and certain physical traits. If you have certain Neanderthal variants, it means that some of your physical traits may trace back to your Neanderthal ancestors.

Variant(s) found

Straight hair
You have 0 Neanderthal variants associated with having straighter hair.
0

Less likely to sneeze after eating dark chocolate
You have 0 Neanderthal variants associated with a reduced tendency to sneeze after eating dark chocolate.
0

Less back hair
You have 0 Neanderthal variants associated with having less back hair.
0

Height
You have 0 Neanderthal variants associated with your height.
0

23andMe tests your Neanderthal variants to sees if you have the specific genes for straight hair, a likelihood to sneeze (or not) after eating dark chocolate, black hair, and height.

Reports

Now that we've wrapped up the DNA Relatives tool, let's look at the rest of the interesting DNA data that 23andMe offers. The Reports>Ancestry tab presents five different sections for you to learn more about your background:

- Ancestry Composition, otherwise known as ethnicity
- Maternal Haplogroup
- Paternal Haplogroup (if the test-taker is male)
- Neanderthal Ancestry
- Your DNA Family

Let's dive into each. We cover the Neanderthal Ancestry report in the sidebar of the same name in this chapter.

Ancestry Composition

Like the other tests we've discussed, 23andMe provides a breakdown of your ethnic heritage as seen in your DNA. Image **E** shows my ethnic breakdown as reported by 23andMe. Because I'm 98.1 percent Ashkenazi Jewish, I don't have much genetic geographic diversity in my reports. Still, the information 23andMe provides is aesthetically pleasing. My results map out where my ancestors came from.

For its Scientific Details section, 23andMe determines ancestry composition using an algorithm that looks at non-overlapping and short DNA segments. Each segment is compared with reference DNA sequences across ancestry populations in 150 countries and territories in Europe, Africa, the Americas, Asia, and Oceania, each made up of thousands of individuals with known ancestry in that region. Your segments are compared with these groups, and your ancestry composition is based on how well you match up with individuals from each reference population. If you share five or more identical DNA segments with individuals from a particular region, that location is assigned to you. Your DNA may match multiple populations in a region, so you'll be given a broader ethnicity (e.g., "Broadly European"). You can view your results in more detail by clicking "See all tested populations" on the bottom of your Ancestry Composition results page.

On the summary page, you'll also be given an estimate of a timeline of your earlier ancestors who don't exhibit a lot of DNA in your results (image **F**). 23andMe is the only company to offer such a timeline, and can be fun to review and share with friends.

Below your timeline is a chromosome browser that shows where on your chromosome you represent different populations. As you can see in image **G**, I exhibit mostly Ashkenazi Jewish (the teal), but also some Broadly European DNA (royal blue) on chromosomes 3, 4, 6, 7, 12, 15, 18, and 20, as well as some Broadly Northwestern European DNA (the light

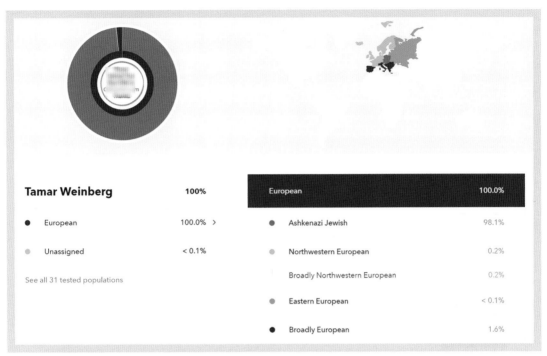

Tamar Weinberg	100%		European	100.0%
● European	100.0% >		● Ashkenazi Jewish	98.1%
● Unassigned	< 0.1%		● Northwestern European	0.2%
See all 31 tested populations			Broadly Northwestern European	0.2%
			● Eastern European	< 0.1%
			● Broadly European	1.6%

Image E. These are my ancestral origins from 23andMe. As you can see, my DNA is almost entirely Ashkenazi Jewish, with some Northwest European, Eastern European, and Broadly European mixed in.

Image F. 23andMe provides a timeline of when, based on your DNA, you had genetic ancestors from a particular region. For example, I have a small percentage of Eastern European DNA that dates back to the 1700s or 1800s. (Your timeline will likely have a wider variety of ancestors.)

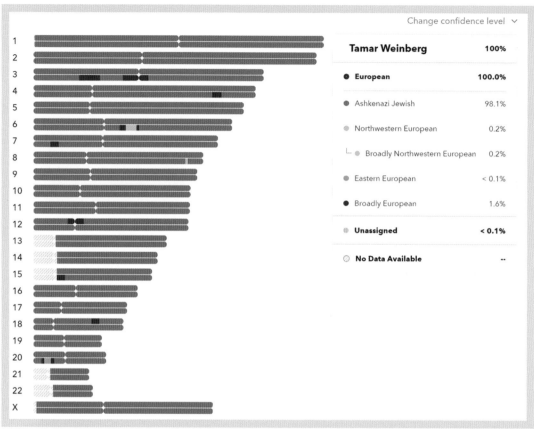

Image G. You can view a chromosome browser to see how much of each region's DNA you have on each chromosome. You can toggle the browser's confidence level at the top right.

blue on chromosome 6) and the Eastern European (the darker teal on chromosome 20). Again, your results will be more exciting than mine—a product of endogamy in my family history. Note that the blue link above my name that says Change Confidence Level. This allows you to change the level of certainty, from 90-percent certainty (Conservative) to 50 percent certainty (Speculative).

Haplogroups

The next section of your results is devoted to your maternal and paternal haplogroups. The details of 23andMe's maternal and paternal haplogroups show the path of migration between Mitochondrial Eve or Y-Chromosomal Adam, plus where your haplogroup resided historically (image **H**). Below that, you can see more detailed (and more interesting) information about the haplogroup, including how common your Y-DNA (if male) and mitochondrial DNA (mtDNA) haplogroups were when they were discovered.

As with haplogroup data from other testing services, you can use haplogroups to help narrow down your list of genetic matches especially if you find maternal and paternal matches. You can also reach out to your fellow members of a haplogroup to investigate each other's family tree (though before you do, refer to chapter 9 on outreach).

You may find 23andMe's haplogroups aren't as frequently maintained or updated as they are on other services. My maternal haplogroup on 23andMe and Family Tree DNA was U4a3, but it changed to U4a3a on Family Tree DNA months ago.

Your DNA Family

This section gives you broad information about the people who are listed in your 23andMe results. Here you can see how many relatives 23andMe determines as close family to second cousins, third to fourth cousins, and fifth to distant cousins. Rather than listing individual results, the Your DNA Family section presents your DNA results in totals: total number of DNA relatives, number of close family to second cousins, number of third to fourth cousins, etc. You can also see a map of the current location of your DNA relatives, where you can toggle between US and world views. 23andMe also provides a

GETTING IN TOUCH WITH 23andME USERS

I've sent quite a few requests to share information with my 23andMe DNA matches, most of them more than a year ago. Responsiveness on 23andMe is a gamble, as people come to the site for their ethnicity estimates or health information, then drop off. They couldn't care less about you trying to find your birth parents or sperm donor. Users now have to attach a name to their profiles, but 23andMe initially allowed users to be anonymous—and many users haven't signed into their accounts in years. That makes getting in touch difficult, so you may have to do without them. You can't imagine how many times (while spacing between months) I've tried to make contacts with my closest anonymous matches.

Keep reaching out to your close DNA matches, though, and don't give up. My first real discovery on 23andMe, my second cousin once removed, needed to be nudged by me about a dozen times over the course of months before he responded. But when he did, he thanked me for hounding him. (He was so excited by his findings that he bought himself an Ancestry.com subscription to learn more about the genealogical roots of our family.) Mind you, most users won't be as forgiving as my relative, but you have to start somewhere. Still, I choose never giving up. You've waited decades for this; you deserve to know.

529andYOU

529andYou is a third-party Google Chrome plugin that helps you view your 23andMe matches in more detail. 529andYou tracks your matches using a local database that resides on your computer. When you first install 529andYou (which you can find in the Chrome Web Store), you will be asked to create a database.

Once you've installed 529andYou, view your DNA match page using Google Chrome. Now, you'll see two buttons near the site's footer: Triangulate into 529andYou and Open 529andYou. When you click on Triangulate into 529andYou, the button will change to Submitting Comparisons. After the comparisons are submitted, you can click Open 529andYou to see the match and their comparisons to you. You can also choose any other person who is either sharing with you or is participating in Open Sharing.

Once you choose a match, you get information that is similar to Family Tree DNA's Chromosome Browser that is quite useful. You can see the chromosome where the match is located, the start point, the end point, the genetic distance (in cM), and the number of single-nucleotide polymorphisms (SNPs) compared in the match.

529andYou offers views like Basic, Profile Links (which links back to the user's 23andMe profile), and Profile Links and Overlapping Segments (most helpful for triangulation, as it will show you people who share the same DNA on the same chromosome segment as you). You can also add an option to Edit Phasing Information and Ancestors, which typically more advanced users use to add a label to their matches and the names of the common ancestors. This tool also becomes useful with another advanced tool called Genome Mate Pro **<www.getgmp.com>**.

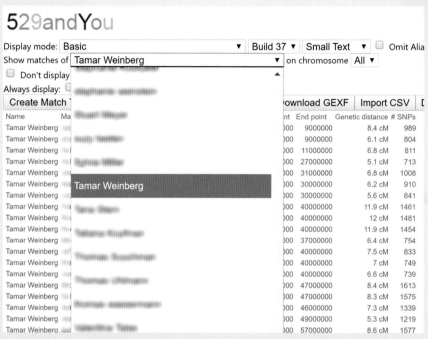

With 529andYou, a Chrome browser plugin for 23andMe, you can choose matches who share DNA with you and run a detailed analysis.

Migrations of Your Maternal Line

L
180,000
Years Ago

L3
65,000
Years Ago

N
59,000
Years Ago

R
57,000
Years Ago

U
47,000
Years Ago

Haplogroup L

If every person living today could trace his or her maternal line back over thousands of generations, all of our lines would meet at a single woman who lived in eastern Africa between 150,000 and 200,000 years ago. Though she was one of perhaps thousands of women alive at the time, only the diverse branches of her haplogroup have survived to today. The story of your maternal line begins with her.

Image H. Like Family Tree DNA, 23andMe provides your haplogroup, as well as a history of your haplogroup throughout the centuries. You can view either a map or a timeline.

chart breaking down the ethnicities of your DNA relatives, plus a graphic showing your matches' likelihood of possessing certain traits compared with the average user.

While not immediately useful for finding genetic relatives, this page can provide you with an at-a-glance summary of your research, plus interesting insights to your genetic tendencies.

8

MyHeritage DNA

yHeritage DNA **<www.myheritage.com/dna>** is the fourth major testing company that we'll cover in this book. Still relatively new to the DNA scene (the test launched in 2016), MyHeritage DNA offers a DNA test that will open your pool of DNA matches up to include even more people. As with other testing services, your results will include an ethnicity estimate and a list of matches in the MyHeritage database. And like Ancestry DNA **<dna.ancestry.com>**, MyHeritage DNA builds off its website's specialty: online family trees and digitized genealogy records.

In early 2018, MyHeritage launched a substantial redesign that boosted the site's usefulness to test-takers. MyHeritage buffed up its matching algorithm and also created a chromosome browser, allowing users to compare DNA results in more detail. Users can also see how they fit into a family tree based on DNA (provided, of course, that you and your match have uploaded one), and the site accepts raw DNA files from AncestryDNA, Family Tree DNA **<www.familytreedna.com>**, and 23andMe **<www.23andme.com>**. This compatibility allows you to compare your test results to those who have tested with other companies, potentially opening you up to even more genetic relatives.

Despite these innovations, some may still find MyHeritage DNA matches to be overrated. The site has a lower threshold for shared DNA than some of its competitors, meaning your relatives on the site may be closer on MyHeritage than they appear in other databases.

Let's dive into the different sections within MyHeritage DNA's toolbox.

Image A. Your basic dashboard shows your ethnicity breakdown as well as some of your cousins listed by closest to furthest.

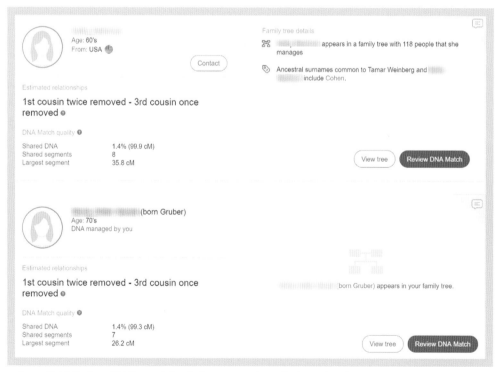

Image B. MyHeritage DNA's match list displays your estimated relationship, the amount of shared DNA, the names of those who appear in both you and your match's tree, and more.

UPLOADING RESULTS FROM OTHER COMPANIES

Before MyHeritage started its own DNA test in 2016, its users had to upload results from other companies if they wanted to make use of MyHeritage's analysis tools. Fortunately, MyHeritage has kept this functionality. You can create a free account, then click DNA from either the site's header or footer. Click Upload DNA, then Start, and answer the necessary questions.

Once you upload your results (which will take twenty-four to forty-eight hours or even longer), you have access to all the DNA tools: ethnicity, DNA match review, and a chromosome browser.

Overview

The first stop under the DNA menu is the Overview section. In it, you'll find an ethnicity breakdown as well as a selection of your DNA matches (image **A**). You can also access ethnicity information under the Ethnicity option in the DNA drop-down menu, and DNA matches through the DNA Matches link.

You can then click on "View 708 DNA Matches" to see a box representing each of your DNA matches (image **B**). The site updates I mentioned earlier really made a difference to my match count: I now have more than seven thousand matches, up from 708 just a few months earlier.

If you manage additional DNA kits, you can select one of the other kits to get the same data: ethnicity and DNA match information. When you first upload the data, you will see a status is still processing message, and it typically takes one to two days (though it may be a little longer). When the results are ready, you can compare them under the other options in the DNA menu. Below the three dots to the right of each test, you can reassign the kit to a different person (in case you mislabeled the kit you uploaded or the person you've ordered the kit for is unable to test), view the result's ethnicity estimate, view DNA matches, download RAW data (assuming you've actually purchased the DNA test through MyHeritage and did not upload your results), and delete your kit.

Ethnicity Estimates

MyHeritage DNA tests more than forty ethnicities throughout the world, an interesting way of helping you learn about you and your match's shared ancestors. Like other testing companies, MyHeritage DNA creates these reference populations made up of people from that region, then compares your DNA to each.

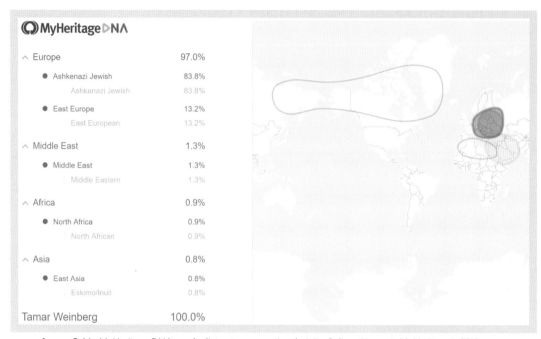

◐ MyHeritage ▷NΛ	
⌃ Europe	97.0%
● Ashkenazi Jewish	83.8%
Ashkenazi Jewish	83.8%
● East Europe	13.2%
East European	13.2%
⌃ Middle East	1.3%
● Middle East	1.3%
Middle Eastern	1.3%
⌃ Africa	0.9%
● North Africa	0.9%
North African	0.9%
⌃ Asia	0.8%
● East Asia	0.8%
Eskimo/Inuit	0.8%
Tamar Weinberg	100.0%

Image C. My MyHeritage DNA results list out my genetic ethnicity. Believe it or not, MyHeritage's 83.8 percent Ashkenazi estimate is actually the most diverse of my results from the major testing companies.

As with other ethnicity estimates, you should take these results with a grain of salt. Ethnicity estimates represent trends in your DNA and are not meant to be precise. You'll likely notice your estimates differ from those at other companies, and may change over time as reference populations and algorithms change. My results, for example, are somewhat different than they are on other sites (image **C**). MyHeritage DNA gives me a small (0.8 percent) amount of Eskimo/Inuit DNA, which I haven't seen in other ethnicity estimates. This oddity might just be a blip as MyHeritage continues to work out kinks in its ethnicity estimates—or perhaps my genetic background really is more interesting than I thought!

As a fun feature, MyHeritage DNA also allows you to view aggregated data about all of its users' ethnicities. Click on the highlighted countries to see the portion of MyHeritage DNA users in a specific country that have that ethnicity (image **D**). You can also toggle to see the full user base by ethnicity. When you hover over any of the ethnicity regions, you can see what ethnicities are associated with that region. You can choose Africa, America, Asia, Europe, Middle East, and Oceania to get the population data.

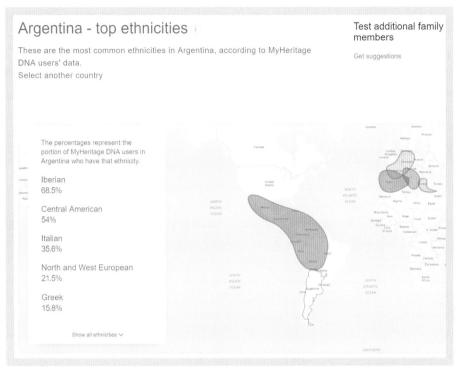

Image D. MyHeritage allows you to view aggregated ethnicity estimates for all of its users. For example, I can see that Iberian, Central American, Italian, North and West European, and Greek are the most popular ethnicities for people who live in Argentina.

RESEARCH TIP

Note Your Privacy

If you're concerned about security or privacy, take comfort in MyHeritage DNA's terms of use. The company does not own, share, or sell your DNA, and you can delete your sample—including the physical sample—at any time. It's your data, and you can do what you want with it. You can even turn matches off at any time to maintain a sense of privacy, though this will obviously cut off your ability to make connections with other users.

Clicking on some of these regions will allow you to see a screen of breaking down each ethnicity within the region. For example, clicking on United States simply shows where Native Americans have emerged, but clicking on Filipino, Indonesian, and Malaysian countries will further break down into Singapore, Thailand, Hong Kong, Netherlands, and India—with percentages for each geographic breakdown. Below that is a paragraph providing some historical context into each region.

DNA Matches

Let's return to your DNA matches. Each result on the page gives you a snapshot of your relationship with that match, including your estimated relationship, the amount of shared DNA and shared segments, and the length of the longest shared DNA segment. The right column also indicates your specific family members who appear in the match's MyHeritage family tree (again, assuming both you and the match have uploaded family trees to the site.)

Click the "Review DNA Match" button for more details (image **E**). Like 23andMe, MyHeritage includes some useful information as it compares your results with your match and the results of matches you both share. You can view shared DNA matches in a table that shows how you and the match are estimated to be related to each person, as well as shared ethnicities.

THE MYHERITAGE FAMILY TREE

MyHeritage is best known for its family tree functionality, and it boasts ninety-one million users with forty-two different language translations. Like Ancestry.com, you can upload a GEDCOM or create a tree from scratch, and with a paid subscription, gain access to records to confirm the paper trail. This is helpful when looking at matches and seeing if there's indeed a known connection.

Creating a family tree is as simple as clicking on the home page (assuming you're not yet logged in) and choosing "Start your family tree." Once you do, you'll be asked to either upload a GEDCOM or start from scratch by providing your name (maiden name if female), year of birth, mother's name, and father's name (if known—if you are adopted, you can use your birth family or adopted family, or both if you create separate trees.)

If you haven't uploaded a GEDCOM, the next page will allow you to input additional relatives: information about your maternal and paternal grandparents (if known). When you're done with this information, you'll have a basic tree. If any special events occur during this time, you may also see a celebratory balloon and more information as seen in this image. You can continue adding family by clicking on the + sign under any name in the tree.

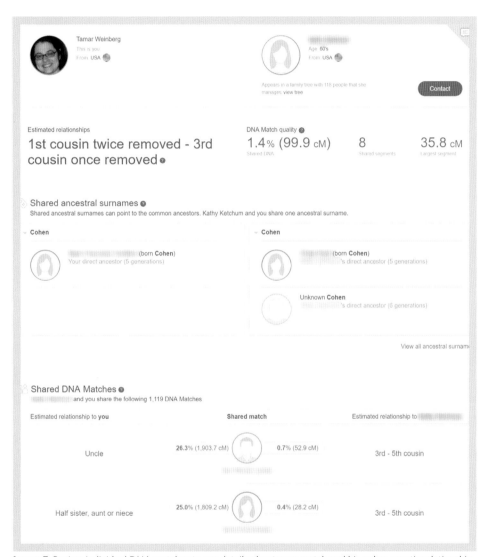

Image E. Review individual DNA matches to see details about your match and his or her genetic relationship with you.

Let's look at the example in the image. My match (we'll call her Jane) shares 1.4 percent of her DNA with me, but she matches my maternal uncle with even less—just 0.7 percent. How can she share less DNA with me than she does with my uncle, who is in an earlier generation? Endogamy is likely to blame here—she may match my maternal uncle's DNA in addition to my paternal grandfather's DNA or even my paternal grandmother's DNA, but likely at far lesser percentages. More advanced tools will help me better understand the relationship. That 99.9 centimorgans (cM) of shared DNA is likely the sum of several small segments, as my largest shared segment with this match is only 35.8 cM. That happens to be a match worth pursuing, so I'm going to explore it later, but I'll need more advanced analysis tools to help me figure out which small segments we share DNA on.

You may also notice from the image that MyHeritage has a shared ancestral surname section. It is possible that we come from the same family (Cohen). I do not know much about my fifth great-grandmother's siblings, but it's a remote possibility that we're related through that name. It's hard to ascertain with certainty, though, because Cohen is a common surname. Again, this will require more investigation.

One of the most exciting things to know about MyHeritage is its dedication to help you find a missing person. According to Rafi Mendelsohn, MyHeritage's public relations director, the process of seeking family is so emotional that some test-takers require handholding. If one contacts MyHeritage for assistance, its employees are happy to go the extra mile and even share stories of positive outcomes on their stories page **<stories.myheritage. com>**, one of which I will share in one of the case studies at the end of this book.

One shining example of this is the DNA Quest project **<www.dnaquest.com>**. Launched in March 2018, DNA Quest is a pro-bono initiative in which MyHeritage donates thousands of DNA kits to US-based adoptees who are seeking biological family (and to parents who may have given children up for adoption). At time of publication, any future plans for the project were still an open question, but the first phase of the project alone (in which more than $1 million-worth of DNA kits were donated for free) shows MyHeritage's commitment to helping adoptees.

Chromosome Browser

MyHeritage DNA's chromosome browser, new in 2018, is exactly what you've come to expect from such a tool. Using the chromosome browser, you can compare a key person (yourself) with up to seven matches to view the DNA segments you all share.

Let's take a look at an example. In image **F**, I use the chromosome browser to compare my DNA with that of my mother, paternal grandmother, paternal grandfather, paternal first cousin twice removed on my maternal grandmother's father's side, and paternal first

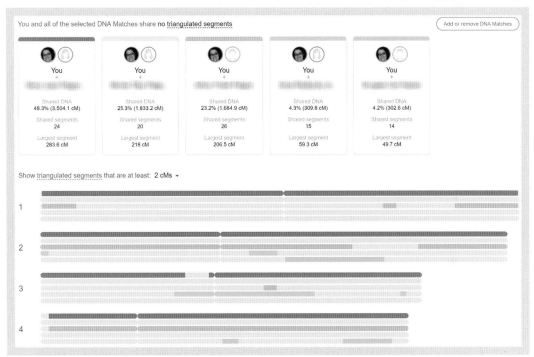

Image F. MyHeritage DNA's chromosome browser allows you to compare DNA segments of up to seven test-takers.

cousin twice removed on my paternal grandfather's mother's side. As expected, I have nearly complete overlap with my mother (the red). Meanwhile, my paternal grandmother (the yellow) and paternal grandfather (the orange) almost never overlap with each other—instead, I share DNA with them each individually, meaning (among other things) that my paternal grandparents were not closely related.

At the same time, my first cousin twice removed on my maternal grandmother's side (the green) overlaps somewhat with my mother on chromosomes 2 and 3—which makes sense, given those two are related by blood. However, that cousin on my mother's side won't match much at all with my cousin on my father's side (the blue) since they're on different sides of my family. Given that, I can confirm they are maternal matches.

On this screen, MyHeritage will also indicate whether you and all the relatives share a common, "triangulated" DNA segment. By design, I have no triangulated segments with this focus—I specifically chose relatives who would not overlap. Normally, I wouldn't bother comparing my DNA to that of my close relatives (like my mother or paternal grandparents). Instead, I'd take myself out of the analysis and compare my relatives' DNA with that of their cousins, since I already know I have significant overlap with my mother and paternal grandparents. However, the illustration helps me learn more about more distant relatives as I know them, and I can use the chromosome browser to determine whether any triangulated segments may have come from a recent ancestor.

What I can definitively say here is that the two maternal cousins (the green and blue) help me determine that our shared ancestors are one of my maternal grandmother's maternal grandparents (green), and the other cousin's most recent common ancestor is one of my maternal grandfather's maternal grandparents (blue). Still, I'd need to test relatives from each of these grandparents to determine which grandparent the DNA truly comes from. I can corroborate as much as I know so far with a paper trail.

When using the chromosome browser, you want to find relatives who have segments that all overlap in at least one area (the definition of triangulation). If you've determined who your parents and grandparents are and they have been tested, you can use those as a baseline to compare the other relatives to—overlapping segments then can be isolated to maternal or paternal sides.

If you do not have direct ancestors just yet, you can still try this with known cousins. However, the overlapping segment you share with others may have come from the other parent, since triangulation requires three test-takers. A known paternal cousin matched to another known paternal cousin as compared to you can be the right recipe for tracing back lineage to their most recent shared ancestor.

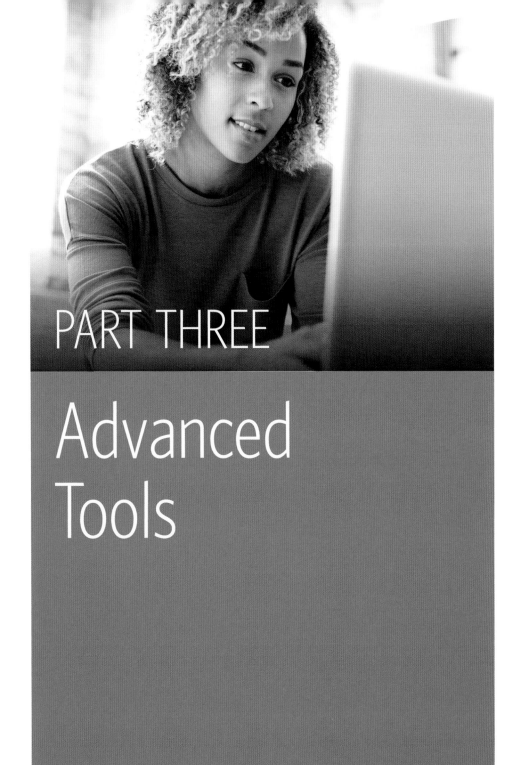

PART THREE

Advanced Tools

9

Establishing a Biological Connection

We've discussed the basics of DNA testing, plus what each DNA test and testing company can offer you. Now all that's left to do is put it all together: How can you connect the long-lost branches of your family tree, and reach out to family members who may have answers? For many adoptees, this can be the most daunting part of the research process.

Some people do their DNA tests just for fun or to learn their ethnicities, unaware of the potential to match with long-lost relatives. Unexpectedly reconnecting with relatives—some of whom were previously unknown or who bring up painful memories—might be outside another user's comfort level, and they might not be as receptive to learning about their past as you are.

Given how delicate these situations are, it's best not to rush into these new relationships blindly or without a plan. In this chapter, we'll discuss two major strategies for connecting with your biological relatives: creating mirror trees and reaching out to matches/potential relatives.

Creating Mirror Trees

Earlier in this book, I mentioned that you shouldn't reach out to another test-taker/potential relative until you've already copied their public family tree. There's always a chance that a match, unable or unwilling to cope with a new relationship, will block your access to their genealogical and genetic information once you reach out and identify yourself as a relative. Though a sad reality, this is more common than you might think—and it can leave the unprepared researcher empty-handed.

That's where mirror trees come in. A mirror tree is an exact replica of another user's tree with one exception: You're the home person. Mirror trees work only on Ancestry.com due to that site's DNA matching capabilities. With a mirror tree, you can use someone else's family information (provided they are a close match) to fill out your family tree in the absence of a direct connection with the other user. Remember to keep your mirror tree private—you don't want other users to think you're stalking them!

Mirror trees are a crucial first step to finding your birth family, particularly in the absence of other information. After all, you share ancestors with one or more of your matches, and mirror trees can help you build out a family tree based on that relationship.

Start by reviewing your match list. AncestryDNA indicates whether or not each of your matches has uploaded a family tree. Though the company suggests all test-takers upload a tree, not all do—and fewer still have made their trees public. Select a match who has uploaded a tree that you can see, then scroll down.

Here, you'll see your match's user-provided family information (image **A**), including a list of the surnames you and your match have in common and a pedigree view of their ancestry (showing parents, grandparents, great-grandparents, and so on). Note that this pedigree view shows only the match's direct-line ancestors and not their siblings, so you may want to add siblings to your mirror tree as well (bearing in mind that a sibling of an ancestor of your match may be the direct maternal/paternal match to you). Obviously, this tree is only as accurate as the user's sources. Still, your closer matches will usually know who their immediate family is.

Click on a name in the tree to view more details about that person. Then click View Full Tree to see that person's complete Ancestry.com family tree (image **B**), which includes siblings and children in addition to direct-line ancestors. Normally, living people

MIRROR TREES AND ENDOGAMY

Mirror trees can be incredibly useful—if you are not from an endogamous culture. If you suspect you're from an endogamous culture, you might consider waiting to build one. Because people from endogamous cultures share an unusual amount of DNA, they have more trouble parsing out which DNA came from whom (and, thus, how their family trees line up). One exception: If you have only a small portion of an endogamous culture's DNA (say, 25 percent Ashkenazi Jewish) you might have an easier time figuring out how that portion fits into your family tree.

Image A. When you view an AncestryDNA match who has uploaded a family tree to the site, you can view surnames you share, as well as the user's family tree.

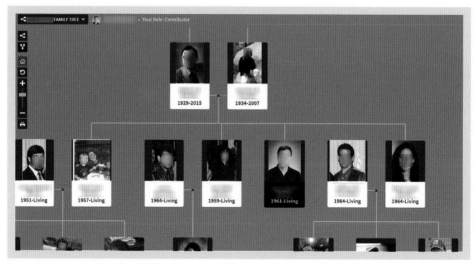

Image B. Be sure to click View Full Tree from a match's page to help you identify potential genetic relatives.

are marked as private, so you may have to work only back with previous generations. But that's probably enough to get you started on a mirror tree. On a piece of paper, start mapping out how the members of your DNA match's family tree connect to each other—and how you think they might connect to you. This will be the basis for your mirror tree.

Once you've done your research, it's time to create your mirror tree by going to Trees>Create & Manage Trees. Make sure you select the option to make your tree private. Then, recreate the family trees you've been researching. Pay special attention to names and birth years to glean insights into who might be a possible parental candidate.

Once you've got relatives charted out, it's time to connect your DNA to your mirror tree. In this structure, you will essentially stand in the place of your DNA match—the person whose relatives you've copied into your mirror tree. Go to your DNA results, then click the Settings tab. Here (image **C**), you'll get a Family Tree Linking option that allows you to pick a certain tree, then find the name within that tree to establish the proper link. The tree will look exactly like your match's, but you will stand in place of the match as the linked person.

AVOID ANCESTRY ERRORS

As we've said before, family trees on Ancestry.com are user-submitted, meaning their quality relies on another user's attention to detail. As a result, verify the accuracy of the information you see before adding data to your mirror tree. Check sources and review the conclusions the other user has drawn. Specifically, look out for these common research errors as you're reviewing others' trees:

- Duplicate records for the same person: For example, having two Jims who were both born in 1847 but died in different years
- Children born after mothers: For example, a child who was born ten years after his mother's death
- Insufficient ages for having children: For example, a parent having a child at just six years old

The list goes on. Learn more about common genealogy mistakes and how to avoid them at *Family Tree Magazine* <**www.familytreemagazine.com/premium/5-common-genealogy-errors**>.

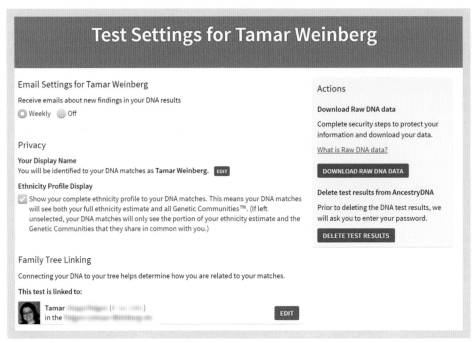

Image C. From your test settings, you can link your results to a person on a family tree.

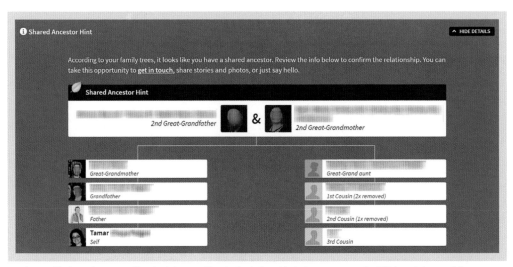

Image D. Ancestry.com will calculate an estimated relationship between you and a DNA match, providing a Shared Ancestor Hint. For example, it predicted that a match and I share my second great-grandfather/mother as our most recent common ancestor.

Once you are properly connected in your mirror tree, Ancestry.com will go to work. The site will scan your tree and match list to identify common ancestors, if any. After a few days, a leaf will show up next to your match—a shared ancestor hint. You can then click View Match and see what the shared ancestor hint looks like (image **D**). The match, as you recall, should be close enough for you find an accurate representation of a shared ancestor. This way, you can find users who match your DNA and have the same people in their tree—your shared ancestor(s). If no hints appear, then that's the wrong ancestral branch.

Your goal should be to find multiple shared ancestor hints that point to the same ancestral branch. And once you've established which branch shares DNA with yours, continue researching that branch. As stated earlier in this section, add spouses, siblings, and children, working forward and backward in time. In time, you'll likely find overlap with other DNA matches who share other ancestors with you. Put these all together by comparing your family trees. Recall that living relatives will be hidden on Ancestry.com, but may be on Facebook or other social media—and don't forget to use obituaries. Some have tremendously helpful information.

WHY NOT USE SHARED MATCHES?

Mirror trees sound like a lot of work! Why can't I just use Ancestry's shared matches feature? The answer is simple: Shared matches don't necessarily provide the right information. In one case, my cousin's mother, Franceska, had two close matches: my mother and a first cousin once removed, Doug. While analyzing the DNA, I saw that the DNA Franceska shared in common with Doug wasn't the same DNA Franceska had in common with my mother. The three definitely appeared to be close cousins, sharing a lot of DNA, but there was absolutely no overlap among the three of them. I wrongfully concluded they may be related, but not the same way. I held onto this theory for a good six months. I called Doug regularly asking for theories. Maybe it was his father's side, not his mother's side that connected him and Franceska. I simply had no idea where to turn. Sometimes, DNA alone doesn't cut it. You may need a paper trail and that mirror tree to help you find what you're looking for.

Comment on Family Trees

Ancestry.com has an internal messaging system, but users who only interface with Ancestry.com through the mobile app can't access it. Instead, you'll have to comment on their public family tree to reach mobile users. To do so, load your DNA match and access their tree. From the Tools dropdown, click View Comments. There, you can find a white box to type a comment. As with other forms of communication, be sure to introduce yourself and include some information about your research and suspected shared ancestor with them.

But sometimes it isn't that easy, and it takes a long, long time—even years. You may not have shared ancestor matches for several reasons. For one thing, your match (the home person you replaced in your mirror tree) might have errors or omissions in their family tree. You may have entered the data incorrectly, or perhaps your match has faulty information—especially common if historical records are no longer available. Perhaps your match has input information about his adopted family, rather than his birth family. Double-check your data and/or start a mirror tree for another ancestor.

What if your strong DNA matches don't have trees? Fortunately, Ancestry.com has other ways of getting in touch with users. A surname search on the site will turn up results from all family trees—even some private ones. Alternatively, you can visit <www.ancestry.com/community> to search the message boards for posts from the individual, or search for the user in the Member Directory (available under the Search menu). Additionally, go to the match's profile page and see if she has another tree listed—some users have both public and private trees. Or perhaps your match simply hasn't linked himself to a family tree (though there are trees for the match), which Ancestry.com translates as "No family tree." From a match's profile page, you can also learn other details useful to reaching her: a biography, additional tests she manages that you may match, and message board posts she's made. You can also use shared matches to identify unknown matches, e-mail your match through Ancestry.com with basic information, or leave a note about the suspected relationship. It's up to you!

If you're not using Ancestry.com for DNA testing, creating a mirror tree is an arduous process. On these other sites, you'll need to manually create your tree using your closest match. If you don't yet know if your match is maternal or paternal, you will need to create different trees until you see some sort of convergence or pattern. This requires process of elimination after you access trees from other family members and see where people match up. The reason Ancestry.com does so well with mirror trees is that, once you add yourself in place of your DNA match, Ancestry.com's system computes possible ancestors from that simple replacement using your other matches.

Reaching Out to Family Members

Most people reach out because they want closure. It's not about money or wanting a share in the late parent's estate—it's about finding out who you really are. It isn't easy to build up the courage to say something that may shake your world.

Outreach is one of the scariest things people of unknown parentage end up having to do. Who wouldn't be terrified of picking up the phone, writing a letter, or sending an e-mail or Facebook message to someone they've never met but might be intimately related to? Is this relative eager to meet you? Has he also been looking? Does he even know you exist? Will you and this other person want to begin a relationship after all these years? You won't ever know until you try.

THE SEARCH ANGELS AND DNA DETECTIVES

Think about the good things that can come from reaching out—the long phone conversations and the new extended families to be discovered. While it doesn't always turn out that way, real stories of other adoptees can help encourage and inspire you to discover more about your family.

Fortunately, a great Facebook community called DNA Detectives **<www.facebook.com/ groups/DNADetectives>** features many of these stories, both positive and negative. Among its members, you'll find people on both sides of the adoption equation who are there to support you. Some of these individuals will volunteer their time to help you find your birth families.

The Search Squad Facebook group **<www.facebook.com/groups/searchhelpers>** can also connect you with "search angels," volunteers dedicated to helping reunite families. This group has more rigorous policies for you to follow (for example, you have to post yourself rather than having a friend help you).

Note that both these communities are volunteer-based. So any offers for paid reunion services on the pages of these groups or of groups with similar-sounding names are inaccurate at best and a fraudulent scam at worst.

I've seen this scenario play out many times throughout my research. Sometimes when I've reached out to the genetic second or third cousins of my clients, the cousins are interested in helping and provide a trail to follow. Sometimes, they won't answer at all, and other times they deny the circumstances leading to my client's birth (such as an unexpected pregnancy). Some of these folks rejoice, ecstatic to find this long-lost relative they also may have spent years searching for. Others send back a "never contact me again" message. Your interactions can create so many mixed emotions—the highest of highs or the lowest of lows. You must be prepared for several different outcomes.

Given this variety in circumstances, you might consider implementing a few strategies in your attempts to reach out to relatives. Perhaps reach out to a friend or relative of your genetic match and ask him to put a word in for you. You can also make your first message far less personal, using a script that begins with something similar to "I'm a genealogist looking into the Brown family." A volunteer from one of the Facebook groups mentioned in chapter 2 may give their time to do that for you as well.

Be sure to save all the information on your biological family before you attempt to reach out, and be careful about the wording of your messages. Keep the lines of communication open, and know that the person may need time to process this information.

Unless you find an exact parent/child match, you probably shouldn't spill the beans right away. Start with your closest matches who are in the third-cousin range or closer (though do bear in mind that not all third-cousin matches know their full family tree and so may not be able to help you). You can build an e-mail that achieves three objectives: introducing yourself, sending a friendly note, and gently requesting information. Here are two possible scripts you can use. The first works better for e-mail, and the second is better if you're calling birth family on the phone.

> Subject: FTDNA Relative
>
> Hello [name],
>
> I notice we match on FTDNA. Our longest segment is **<#>** cM. I was hoping we can compare our trees to determine who our common ancestor may be. I've attached a chart that might help us out. I hope to hear back from you soon.
>
> Sincerely,
>
> [Your name]
>
> [Your FTDNA name] on FTDNA

Another outreach attempt would be a script used during a phone call and might sound like this:

My name is [name], and you appear to be a DNA match. I'm looking to fill in missing parts of my family tree, and it appears we may be connected through the [surname] family line. I was born on [date], in [location]. Does this sound familiar at all to you?

This is the most subtle way to reach out to a relative, and it usually works if the family member is interested in collaborating. Regardless of how you reach out, take comfort in knowing that, no matter how your potential relative responds, you've come much closer to the closure you seek.

Your first attempt at contact should not include the word "adoption." That word can easily scare off potential relatives, unless, of course, they acknowledge in their public profile somewhere that they want to help adoptees. For example, in GEDmatch <www. gedmatch.com> (see chapter 10), you may see matches with DD that prefix their alias, such as "DD Tamar." DD means they are members of DNA Detectives and would be interested assisting adoptees find matches. Regardless, use the earlier script or a variation to guide you into getting family information based on what you already know.

When preparing to reach out to a match you believe will be more open, tell your relative what you know. For example, one of my cousins tested and found out his birth father wasn't the man who raised him—a non-paternity event (NPE). We knew his birth date and where his father would have been. With that information, we narrowed the results to three brothers. One of them was married, so we assumed it was one of the remaining two. Fortunately, the father had tested his DNA for fun, and (since he and I happened to already know each other) I had him upload his data to GEDmatch (chapter 10) to compare with the son. The results confirmed my hunch, and I got on the phone with him immediately and told him that he has a son. He'd had no idea.

Normally, you should make the initial contact—unless you have a relative who is more closely connected and would be able to help connect the dots for you. Remember to use the script! If there's a positive match based on the birth detail you provide, you may be asked if you're adopted. If so, share more information from your birth certificate, such as the name you were given at birth, the hospital you were born in, or non-identifying information such as "I believe my mother was a librarian at the time."

If your contact was well-received, congratulations—you did it! If not, it may take some time for the family member to process the news. Be patient and hope that they come around.

If you don't receive a reply, one of several things could be at work. First, some people are concerned only with ethnicity and not interested in the DNA match element of these test-taking services. They may log in once, then never again. In a similar vein, more and more people are having their DNA tested—many of them just people who are interested in their ethnicity and/or the family tree, not genealogists. But they may not have many details about their family tree. Finally, your messaging may overwhelm the other user if it's overly vague, not simple, or rambling.

Remember, you're going to have to put the pieces together. It takes effort, hard work, and time. But hopefully the guidelines in this chapter will help you reach out to close matches or your parent/child and reunite.

10

Analyzing Your DNA with GEDmatch

Now we're getting to the most exciting part of DNA testing. Though we've spent a lot of time covering the big DNA testing companies, many genealogists consider GEDmatch <www.gedmatch.com> to be the best site for doing deep DNA analysis. With its comprehensive analysis tools, GEDmatch can provide you with incredibly useful insight for tracking down your birth family and building out your family tree.

You can't order a DNA test from GEDmatch. Instead, the free tool accepts raw DNA data from the major DNA testing companies we've discussed in this book: 23andMe <www.23andme.com>, AncestryDNA <dna.ancestry.com>, MyHeritage DNA <www.myheritage.com/dna>, and Family Tree DNA <www.familytreedna.com>. (GEDmatch is also compatible with data from GenetiConcept <geneticoncept.com> and WeGene <www.wegene.com/en>, testing services we have not discussed in this book.)

This chapter will teach you all you need to know about GEDmatch. Like the other services covered in this book, GEDmatch may undergo some changes between this book's publication and when you read it. For example, at time of publication, GEDmatch is building a new site called Genesis that will ultimately offer the same features, plus more. However, the basic principles of the site's analysis tool will remain the same. Download your raw DNA from your preferred testing service, and let's begin! (See the Step One: Download Your Raw DNA Data sidebar for more on how to do this.)

Getting Started

To begin, create your free account on GEDmatch. You'll need just an e-mail address, your name (plus an alias, if you prefer more privacy), and a password. (Again, the service is free.) Then, upload your DNA by clicking on Generic Upload *FAST* (under File Uploads). Fill out the information on the page that follows, including the name of the donor, an alias (if you prefer more privacy), the gender of the donor, and the mitochondrial DNA (mtDNA) haplogroup or Y-chromosome (Y-DNA) haplogroup (if available). Then agree to an acknowledgment that you have permission to upload this DNA sample and choose your raw data to upload to the system.

Once you've uploaded your results, you'll receive a kit number, which helps you track your data and can be given to other users. Your kit number will begin with a letter indicating the testing company. This is useful when comparing your results with those of another GEDmatch user, as the different testing companies use different algorithms. The options for the DNA testing services we've covered in this book are:

- A for AncestryDNA
- H for MyHeritage DNA
- M for 23andMe
- T for Family Tree DNA
- Z for kits with an unknown service

This kit number (which you can find on your dashboard at any time) unlocks a whole world of DNA genealogy tools for you. Beginning immediately with your upload, you can access the One-to-One Matches tool, though you'll have to wait between twelve and twenty-four hours to access other tools.

Comparison Tools

Let's now turn our attention to each of the analysis tools, beginning with tools that compare your results to those of other users and (hopefully) help connect you to genetic relatives.

'One-to-Many' Matches

As with the major testing companies, GEDmatch compares your results to those of its other users. GEDmatch's One-to-Many Matches tool, however, provides much more detail about how your data lines up with others. And unlike AncestryDNA and 23andMe (which do not accept uploads from other services), GEDmatch test-takers have tested at a variety of companies, allowing you to compare your data against a wider pool of users, including Family Tree DNA and MyHeritage DNA. You can see the results of my One-to-Many analysis in image **A**.

Kit Nbr	Type	List	Select	Sex	GED/WikiTree	Haplogroup		Details	Autosomal			Details	X-DNA		Name (* => alias)	Email
						Mt	Y		Total cM	largest cM	Gen		Total cM	largest cM		
A	F2	L		F		U4a3a		Δ	3587.1	281.5	1.0	X	196	196	Tamar Weinberg	@gmail.com
A	F2	L		M			J1a3a	Δ	3587.1	281.5	1.0	X	196	196		@gmail.com
M	V4	L		F		U4a3a		Δ	3587.1	281.5	1.0	X	196	196		@gmail.com
A	F2	L		F		U4a3a		Δ	3586.9	281.5	1.0	X	196	196		@gmail.com
A	F2	L		M		N1b1b1	R-Y19847	Δ	3586.9	281.5	1.0	X	196	196		@gmail.com
A	F2	L		M		U4a3a		Δ	1961.8	131.2	1.4	X	28.9	23.2		@gmail.com
A	F2	L		M		U4a3a		Δ	1897.1	113.9	1.5	X	30.5	24.9		@gmail.com
A	F2	L		F		N1b1b1		Δ	1869.9	218.6	1.5	X	196	196		@gmail.com
A	F2	L		M			R-Y19847	Δ	1735.9	214.5	1.5	X	0	0		@gmail.com
A	F2	L		F				Δ	812.2	115.2	2.1	X	0	0		@gmail.com
A	F2	L		M				Δ	768.6	95.5	2.1	X	0	0		@gmail.com
A	F2	L		F				Δ	643	78.9	2.2	X	0	0		@gmail.com
A	F2	L		M				Δ	388.4	48.9	2.6	X	0	0		@gmail.com
A	F2	L		M				Δ	340.2	48.6	2.7	X	36.7	27.1		@gmail.com
T	F2	L		F				Δ	311	56.0	2.8	X	50	43.7		@gmail.com
M	V4	L		F		K2a2a		Δ	261.4	54.5	2.9	X	23.3	23.3		@yahoo.com
A	F2	L		M				Δ	253.4	63.9	2.9	X	0	0		@gmail.com
M	V3	L		M		H6a1a1a	R1b1b2	Δ	225.5	59.2	3.0	X	8.4	8.4		@gmail.com
A	F2	L		F		U4a3a		Δ	212.5	27.3	3.0	X	42.7	42.7		@gmail.com
A	F2	L		F				Δ	211.4	33.9	3.0	X	34.4	19.6		@gmail.com
T	F2	L		M				Δ	168.2	34.5	3.2	X	0	0		
A	F2	L		F				Δ	163.4	88.5	3.2	X	0	0		@gmail.com
M	V3	L		F		K2a2a		Δ	148.4	37.1	3.3	X	0	0		@yahoo.com
M	V3	L		F		K2a2a		Δ	146.5	37.1	3.3	X	0	0		@gmail.com
T	F2	L		M				Δ	142.8	22.6	3.3	X	0	0		@gmail.com
A	F2	L		M				Δ	139	33.7	3.3	X	0	0		@aol.com
A	F2	L		M				Δ	126.8	57.1	3.4	X	6.6	6.6		@gmail.com

Image A. GEDmatch's One-to-Many feature compares a selected kit to all the tests in the database, listing two thousand that are closest matching.

My results include my son, myself, my parents, grandparents, uncles and aunts, first cousins once removed, and even a second cousin once removed. But what is it that we're actually looking at? This is often the most confusing part—a lot of words and no visuals. From left to right, here's what each column represents:

- **Kit Number**: The kit number is the number assigned to other people who have uploaded their autosomal DNA results to GEDmatch, using the naming convention established earlier: a letter indicating the test provider (or lack thereof), followed by several numbers. Notice my results includes As, Ts, and Ms—test-takers from AncestryDNA, Family Tree DNA, and 23andMe.

- **Type**: This represents the type of chip used in the data that was uploaded. The significance of this is outside the scope of this book, but make a note of it anyway in case you notice a discrepancy in your results.

- **List**: Click the L to view a One-to-Many analysis for the other user.

- **Select**: Click this box next to three or more users, then hit submit on the top of the page to compare the data from just the selected users.

- **Sex**: As the name implies, this indicates the test-taker's gender (male, female, or unknown).

- **GED/WikiTree**: If your match has uploaded a GEDCOM or information from WikiTree <www.wikitree.com>, you'll see it here.

STEP ONE: DOWNLOAD YOUR RAW DNA DATA

No matter where you test, you'll likely want to download your raw data so you can use other tools (such as GEDmatch) to further analyze your DNA. In addition to third-party sites, your raw DNA data is also compatible with certain other testing services, including Family Tree DNA and MyHeritage DNA. See the chapters on those tests for more about this.

Here's how to access and download your raw data from each of the major testing companies. Note: Don't open your raw DNA data file, as there's a decent chance you'll mess something up and render it useless.

AncestryDNA

Download your AncestryDNA raw data from your DNA test settings page, where you'll see a link for Download Raw DNA Data. You will need to input your password when prompted, then agree to a disclaimer that your data, once off the site, will no longer be protected under Ancestry's privacy measures. All sites, in fact, use similar verbiage. Don't fret—just click on the Confirm button. Shortly thereafter, you'll receive an e-mail where you can click a one-use link to confirm the download.

I highly suggest saving your Ancestry file, which may default to the filename AncestryDNA.txt, with your name and date. Therefore, I'd call my file tamar-weinberg-AncestryDNA-061318.txt. This becomes particularly helpful if you end up testing other relatives and you become a manager of their accounts, which gives you access to download their DNA data.

> **Actions**
>
> **Download Raw DNA data**
>
> Complete security steps to protect your information and download your data.
>
> What is Raw DNA data?
>
> **DOWNLOAD RAW DNA DATA**
>
> **Delete test results from AncestryDNA**
>
> Prior to deleting the DNA test results, we will ask you to enter your password.
>
> **DELETE TEST RESULTS**

AncestryDNA Download Raw DNA Data link, under DNA test settings

Family Tree DNA

Downloading your raw data on Family Tree DNA is a bit more challenging. You can find the orange Download Raw Data link on your dashboard. If you cannot see it (it's on the bottom of the image that follows), you can also do a Command-F or Control-F to search for the phrase "raw data." You'll get a similar disclaimer about privacy, then you can scroll down the page to download one of six types of raw data. For our purposes, we want the bottom right, Build 36 Concatenated.

Family Tree DNA Download Raw Data link, available from the myDashboard page

23andMe

Get your raw 23andMe data under the Tools menu, then Browse Raw Data. Click this, then click the little blue download link on the next page. You'll then learn what the file contains and view a few messages, including one that says the utility of raw data might be of "limited practical usefulness." (The serious researchers among us would be inclined to disagree!) On the bottom of that page, keep the default settings and download "All DNA," then enter your password to continue. You'll be sent an e-mail saying your download is ready shortly thereafter.

MyHeritage DNA

If you've tested with MyHeritage DNA, go to the Manage Kits section, click the three dots next to your kit, and you will see an option to download raw data.

TOOLS	RESEARCH
All Tools	
DNA Relatives	
GrandTree	
Your Connections	
Share and Compare	
Find Genetic Counseling	
Forums	
Browse Raw Data	

23andMe Browse Raw Data link, under the main toolbar's Tools tab

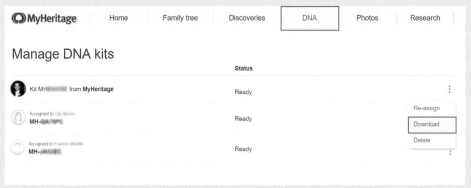

MyHeritage DNA download link, available under the Manage DNA kits menu

- **mtDNA haplogroup** and **Y-DNA haplogroup**: This refers to the haplogroup you may have been issued by tests that provide mtDNA and Y-DNA haplogroups (such as 23andMe's test or Family Tree DNA's mtDNA or Y-DNA tests). Note that users have to enter this information manually, and as such it may be inaccurate (if it's present at all). Some users, confused by this column, might have entered other information here, such as ethnicity.

- **Autosomal Details**: Click on the A icon to see a more detailed comparison between the match and your kit.

- **Total cM** and **Largest cM**: This indicates total amount of centimorgans (cM) and the largest shared segment shared, respectively. The higher the amount of shared cM (and the larger the longest shared segment), the closer the relationship. Use the charts and interactive tools in chapter 4 to estimate the potential genetic relationship.

- **Autosomal Gen**: While an estimate, this represents the generational distance GEDmatch estimates is between you and your match's most recent common ancestor. A distance of 1.0 is parent/child, while a 1.1 to 1.9 indicates aunts, uncles, or grandparents. Larger distances, such as 2.0 to 3.0, are usually first cousins once removed, first cousins twice removed, great-aunts or great-uncles, etc.

- **X-DNA Details**: Click the X to compare your X-chromosome (X-DNA) with that of your match.

- **Total cM** and **Largest cM**: Similar to the autosomal DNA columns, these indicate the total number of cM shared between you and the match and the largest shared segment of X-DNA, respectively. Again, the more shared X-DNA and the larger the shared segment of X-DNA, the closer the relationship. We'll discuss this again later, but note you'll want to run a separate One-to-One DNA comparison on X-DNA matches, as the numbers are often misreported. The One-to-Many algorithm processes up to two thousand results quickly, while a One-to-One test has a more thorough algorithm.

The table is rounded out by the test-taker's name and e-mail address. If a name is preceded by an asterisk (*), the name is an alias. (Though, again, a user may have misunderstood this column as well.)

So what can you take away from this analysis? The One-to-Many analysis can give your research on GEDmatch a starting point. These users could be your first point of contact in your ancestor search, and the quick links to more detailed analysis tools can help you dive further into your possible genetic connections with these folks. In addition, as briefly discussed earlier, the Autosomal Gen category can provide useful hints about your estimate relationship to each user. Here's a breakdown of what the different values may represent:

Estimated number of generations to most-recent common ancestor	Estimated relationship*	Estimated most recent common ancestor
1.0	parent/child	n/a
1.2	sibling	parents
1.5	half sibling, aunt/uncle/niece/ nephew, grandparent/grandchild	grandparents
2.0	first cousin	grandparents
2.5	first cousin once removed	grandparents, great-grandparents
3.0	second cousin	great-grandparents
3.5	second cousin once removed	great-grandparents, great-great- grandparents
4.0	third cousin	great-great-grandparents
4.5	third cousin once removed	great-great-grandparents, great-great- great-grandparents
5.0	fourth cousin	great-great-great-grandparents
*This list of relationships is not exhaustive. Other combinations are possible.		

If you are from an endogamous community, you may want to raise the threshold to a higher number to get more accurate matches, as the One-to-Many analysis will likely overestimate your relationship to other users. A 20-cM segment size would be ideal. This will return fewer matches, but the matches you do receive will be more reliable.

A word of caution: The One-to-Many analysis is not as thorough as you'd expect, and you may notice some inconsistencies in this data as compared to results from the One-to-One Tool (see the next section). For example, my grandmother and one of her matches from this analysis are stated to share over 100 cM on the One-to-Many tool. However, the One-to-One Compare tool shows only a 59.9 cM overlap. It's still a match, but not as close as anticipated. (For comparison, AncestryDNA also overestimated this relationship.)

As stated earlier, this is especially true for the data in the X-DNA columns, because a 20-cM One-to-Many match may turn into a 0-cM match on the One-to-One. Make sure to use the One-to-Many tool in conjunction with the two tools that follow, because they will give you the accuracy sought to truly determine a potential relationship.

One-to-One Compare

The next important feature is called One-to-One Compare, which (as the name implies) compares your DNA kit to just one other person's, providing a more specific and detailed analysis. You can access this tool either from your GEDmatch home page or by clicking on

Raise the Thresholds

Though (in most cases) you'll want to leave the default GEDmatch settings in place, you may want to raise the thresholds when running the One-to-Many and One-to-One tools if your relatives came from an endogamous culture. Since your endogamous genetic relatives on average share more DNA with each other than typical, raising the segment threshold to at least 20 cM or 25 cM can help you weed out untraceable relatives.

one of the A links on your One-to-Many page. If you access it via the former, you'll need to enter the other user's kit number. If you access it from the latter, the kit numbers will be filled in for you. You'll most likely want to keep the default settings for the comparison, as changing them (e.g., by lowering the minimum segment to be included) can lead to too many resulting segments, many of which could be meaningless. If endogamy is present in your family, for example, you'll want to raise the threshold to 20 cM or more as smaller segments may be from a different, more distant ancestor.

Once GEDmatch runs the analysis, your results will look something like those in image **B**, which shows how my kit compares with that of a genetic relative. Your results will indicate what specific chromosomes you and the other user share DNA on, plus the start and stop locations for that shared DNA, the number of shared cMs, and the number of shared SNPs. These outputs will be extremely helpful for triangulation, so save this data or keep it in mind for later. (We'll explore triangulation in the next chapter.)

This user and I have two significant matches on chromosomes 2 and 8, a match of moderate significance on chromosome 5, and a tiny match (which could be anything, considering my family's circumstances of endogamy) on chromosome 11. The largest segment is highlighted, as well as the total number of segments larger than 7 cM. Because of that big 88.4 cM segment, GEDmatch estimates the number of generations to our most recent common ancestor is 3.3 generations (roughly a second cousin, second cousin once removed, or first cousin three times removed). From previous research, I know she's my first cousin three times removed—so not bad!

As with the One-to-Many analysis, your results follow a fairly straightforward pattern: The larger the segment is, the more promising it is to find a real relative match, especially if you share more than 100 or 150 cM (again, with the exception of those in endogamous families). And as stated earlier, the One-to-One comparison is far more precise than the

Comparing Kit T██████ (Tamar Weinberg) and A██████ (██████)

Minimum threshold size to be included in total = 500 SNPs
Mismatch-bunching Limit = 250 SNPs
Minimum segment cM to be included in total = 7.0 cM

Chr	Start Location	End Location	Centimorgans (cM)	SNPs
2	195,701,551	228,377,979	38.8	4,302
5	173,515,657	180,615,468	14.4	1,327
8	28,671,352	129,076,145	88.4	11,387
11	112,621,581	117,986,641	8.9	1,147

Largest segment = 88.4 cM
Total of segments > 7 cM = 150.6 cM
4 matching segments

Image B. Like the big commercial DNA testing companies, GEDmatch can compare your DNA kit to that of another user. This will provide detailed information about the chromosome you share DNA on, plus the start and end locations of the shared DNA and the number of shared DNA (in terms of both cM and single nucleotide polymorphisms, or SNPs).

One-to-Many analysis, so you should definitely consider running a One-to-One comparison to verify all matches. This is also why GEDmatch has a warning in red above your One-to-Many matches that states you should verify relationships with the One-to-One tool.

X One-to-One

Below the One-to-One Compare is the X One-to-One feature. Like the (autosomal) One-to-One analysis, this tool compares your DNA (in this case, your X-DNA) to that of another user. Also like the One-to-One tool, you simply need to enter the other user's kit number and input threshold settings (unless you click the X under Details on the One-to-Many chart). The X One-to-One can be especially useful when male relatives are analyzed, since men have only one X chromosome from their mother, thereby confirming their X-DNA matches are through their maternal lines. If you're a male who matches with a male on the X chromosome, you can definitively say that both of your mothers are related. If you are a female who X-DNA matches a male, however, you may be related through either parent.

Consider this example: My X One-to-One results between me and my male third cousin once removed are interesting (image **C**). You wouldn't think that (given our distant relationship) we share a meaningful amount of X-DNA, but we do thanks to X-DNA's unique inheritance pattern (see chapter 4). In both of our lines, we were able to

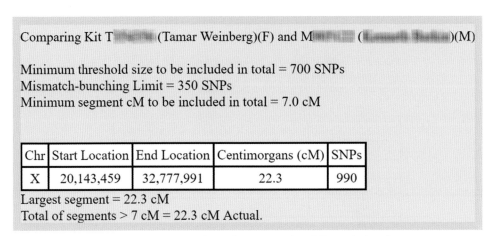

Comparing Kit T██████ (Tamar Weinberg)(F) and M███████ (██████ ██████)(M)

Minimum threshold size to be included in total = 700 SNPs
Mismatch-bunching Limit = 350 SNPs
Minimum segment cM to be included in total = 7.0 cM

Chr	Start Location	End Location	Centimorgans (cM)	SNPs
X	20,143,459	32,777,991	22.3	990

Largest segment = 22.3 cM
Total of segments > 7 cM = 22.3 cM Actual.

Image C. In addition to comparing autosomal DNA between two kits, GEDmatch also allows you to compare two kits' X-DNA.

inherit the X-DNA throughout the generations. His great-grandmother was a sister to my great-great-grandmother. Her X chromosome was passed to her son, who passed it to a daughter, who then passed it onto him. My great-great-grandmother had a son who had a daughter who had a son: my father. In both instances, the X chromosome is making it down through the generations, and that is why we share that 22.3 cM segment.

Having said that, the X One-to-One can be tricky. Many people have trouble finding maternal ancestors, and X-DNA recombines less often than autosomal DNA. As a result, you won't necessarily be able to trace even a large X-DNA match without a paper trail. You also shouldn't trust X-DNA matches in the absence of other data, so always use the X One-to-One in conjunction with the general One-to-One match. Autosomal DNA matches (e.g., from the One-to-One match) are of greater significance than X-DNA matches, and you should always use the latter with the former.

In a previous section, we talked about how the accuracy of the One-to-Many data compared to One-to-One. In one example, the One-to-Many says that I share 20.7 cM of X-DNA with the comparison kit. But in reality, when viewing the X One-to-One, we see that we share nothing. Remember to verify! Always do One-to-One comparisons both on the autosomes and on the X chromosome.

Phasing

The next GEDmatch feature is a tool called Phasing, which is particularly useful for adoptees who know and have tested one of two parents. In short, phasing is the process of figuring out which DNA you received from which parent. By comparing a child's kit to that of one of his parents, you can determine what DNA the child received from the other parent. For example, if you have tested yourself as well as your mother (but not your father), you can compare your results and draw some conclusions about your father's DNA. Whatever DNA you and your mother don't share (i.e., that you didn't receive from your mother) must have come from your father.

GEDmatch can help you with this by creating two artificial "phased kits" that isolate the DNA inherited from your mother and father. Because your parent has given you 50 percent of their DNA, you end up with two kits—each with half of your DNA that you received from your mother and father. These artificial kits don't appear in One-to-Many results, but you can then use them to infer what DNA matches you share with your other parent. For example, you could get a kit that represents your mother and you combined (with a kit number labeled, for example, *PA123456M1* for Phased Ancestry [kit number] Maternal and *PA123456* for Phased Ancestry [kitnumber] Paternal). Run this artificial kit in the One-to-Many tool and compare the phased kit's results with those of your own. The users who match your mother's phased kit are maternal matches. Likewise, if you have only your father's DNA information, you can create a phased kit and run it through One-to-Many to understand what matches you and your father share.

People who match one or both of 2 kits

This tool with a long-winded title can be another great tool for searching for your genetic relatives, and it's especially useful in conducting basic triangulation. Using this feature, you can analyze two different kits and attempt to find a third match that is in common with both.

People who match both kits, or 1 of 2 kits

Kit 1: T▓▓▓▓ (Tamar Weinberg)
Kit 2: T▓▓▓ (▓▓▓▓▓▓▓)

Submit   Select 2 or more from 'Select' column, and click this button for additional display and processing options.

| Match | T▓▓▓▓ | | | T▓▓▓▓ | | | Generations Difference | Email | Select |
	Shared	Largest	Gen	Shared	Largest	Gen			
A▓▓▓▓	3587.1	281.5	1.0	168.2	34.5	3.2	2.2	▓▓▓▓▓@gmail.com	⬚
A▓▓▓▓	3587.1	281.5	1.0	83.7	19.7	3.7	2.7	▓▓▓▓▓@gmail.com	⬚
M▓▓▓▓	3587.1	281.5	1.0	152.4	31.6	3.3	2.3	▓▓▓▓▓@gmail.com	⬚
A▓▓▓▓	3586.9	281.5	1.0	376.0	51.1	2.6	1.6	▓▓▓▓▓@gmail.com	⬚
A▓▓▓▓	3586.9	281.5	1.0	27.8	10.1	4.5	3.5	▓▓▓▓▓@gmail.com	⬚
A▓▓▓▓	1961.8	131.2	1.4	39.8	10.1	4.2	2.8	▓▓▓▓▓@gmail.com	⬚
A▓▓▓▓	1897.1	113.9	1.5	29.0	10.7	4.5	3.0	▓▓▓▓▓@gmail.com	⬚

Image D. The "People who match both kits, or 1 of 2 kits" test uncovers kits that other users whose DNA overlaps with both the selected kits, and this table shows the results between my grandfather's first cousin and me. The test also produces separate tables (not pictured) of users who match one—but not both—of the kits.

Let's assume you have two known relatives who share the same ancestor. By using this tool, you can find users who match you both, followed by users who match only one of the two kits. If someone is flagged as sharing DNA with both you and a relative on the same chromosome and segment, you almost certainly have a recent common ancestor connecting the dots. But if another user matches you both on different segments and chromosomes, you might also share a common ancestor—or you just matched by chance. You'll have to do further analysis to determine which is the case.

You can see my results in image **D**. As you can see in the top row, we're comparing two kits: mine and that of my paternal grandfather's paternal first cousin (which has been blurred for privacy reasons). The table in the image lists the DNA kits that match both of us. The first column lists the name of the matching kit number, while the columns that follow compare the matching kit to my kit (left) and my grandfather's cousin's kit (right) in a handful of categories:

- **Shared cM**: The total number of cM shared
- **Largest cM**: The largest segment of shared DNA
- **Gen**: The number of generations between the matches and each kit
- **Generations Difference**: The generational difference value between the two compared kits (kit 1 minus kit 2, or kit 2 minus kit 1) as compared to the third kit.

These results illustrate the variation of matches between me and my grandfather's cousin, and these variations can help you determine how you, your matches, and this other kit relate to each other. The first result is one of my other kits, followed by my son's kit (second row). Notice how my kit (in the second-to-left column) has higher shared cM values than my grandfather's cousin's kit (in the third column). This is because my grandfather's first cousin (my first cousin twice removed) is a more distant match to my son than he is to me. As a result, my son and I share much more DNA with each other than either of us do to him.

The caveat about this tool is that, with its identified defaults, it doesn't always work precisely with endogamous cultures, especially my Ashkenazi Jewish roots. For example, my first cousin twice removed (my paternal grandfather's first cousin) shares DNA with

RESEARCH TIP

Run Diagnostics

Worried your DNA upload was corrupted? You may want to run diagnostics to make sure everything's alright. Click DNA File Diagnostic Utility from the home page, then enter your GEDmatch kit number. You'll will receive a report of how many SNPs have been uploaded per chromosome, whether the kit is ready for One-to-One comparison, and other statistics. If there are problems with the kit, you'll be informed, or otherwise you'll see "No problems are found with this kit."

her because of where we came from, not because of a shared common ancestor in a genealogically useful time frame. Instead of using the default number, then, people from endogamous cultures should consider a lower number, like 3 or 4, to get more relevant matches.

3D Chromosome Browser

Among GEDmatch's most versatile tools is its 3D Chromosome Browser, which examines three to ten kits simultaneously and identifies shared matches between these selected individuals. It allows you to see how matches are related to each other and which segments they all have in common, helping you to potentially identify a single common ancestor for all of the matches. Results come in the form of a matrix that identifies what segments you all share in common, the total amount of shared cM, and the total shared X-DNA, as well as summary by chromosome.

The output of the 3D chromosome browser first begins with a few charts that display segments in common, total shared cM on the autosomal and X chromosomes (separately), a summary by chromosome, and segment details. My results (image **E**) compare

cM color coding	< 3 cM	3 - 5 cM	5 - 10 cM	10 - 20 cM	20 - 50 cM	50 - 100 cM	Over 100 cM

Segments in common:

Kit	Name	A	T	A	M	Tot. Segments	Largest cM
A		-	41	66	7	114	209.3
T		41	-	35	10	86	106.2
A		66	35	-	5	106	209.3
M		7	10	5	-	22	62.3

Total Shared cM (Chr 1-22):

Kit	Name	A	T	A	M
A		-	996.5	2617.6	140.5
T		996.5	-	1079.8	222.3
A		2617.6	1079.8	-	50.7
M		140.5	222.3	50.7	-

Total Shared cM (X-Chr):

Kit	Name	A	T	A	M
A		-	6.7	123.2	None
T		6.7	-	6.5	None
A		123.2	6.5	-	None
M		None	None	None	-

Summary by Chromosome:

Chr	Tot. Matching Segments for all individuals.	B36 Graphic Posn Range From	To	Largest segment
1	11	72017	247185615	106.2
2	12	8674	242697433	58.8
3	9	36495	199322659	97.5
4	13	49009	191200760	115.0
5	5	78452	180630744	209.3
6	5	100815	170761395	125.8

Image E. GEDmatch's 3D Chromosome Browser gives you a detailed look at how up to ten different kits compare on each of the twenty-two autosomal chromosomes.

four of my paternal relatives: my grandfather, his paternal first cousin (my first cousin twice removed), his sister (my great-aunt), and his second cousin once removed (my third cousin once removed). Each chart, though evaluating different factors, is set up in the same way. Each person is listed in the same order on the rows as in the column, and the white dashes indicate they are the same person at that value and so no comparison is needed.

The first chart shows the number of segments each of the four test-takers have in common. Naturally, the second cousin once removed (who is listed fourth, at the bottom of the table and in the right-most column) is going to share fewer segments than the other three (who are all siblings or first cousins to each other).

The second chart displays total shared cM. The second cousin once removed, as you can see, shares significantly different DNA with her relatives, between 50.7 and 222.3 cM. A shared cM value of 222.3 is high for cousins this distantly related, but still within the "second cousin once removed" range according to genetic genealogists.

The third chart also measures shared cM, albeit on the X chromosome. As you can see, the second cousin once removed does not match any X-DNA with the other three. The reason is that her relationship to the others is almost strictly paternal. While she inherited an X chromosome from both her father and mother, her father's father is our direct relative, as is her father's father's father. (Remember that fathers never give the X chromosome to their sons; they pass down the Y chromosome to sons.) As such, she would not be expected to have an X chromosome match since the relationship was strictly on the paternal line until it ended with her.

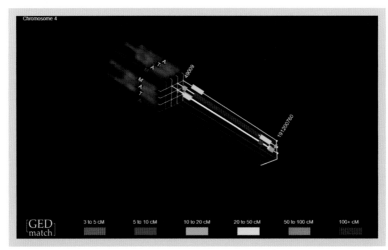

Image F. You can view the results of the 3D Chromosome analysis either as a table or as a graphic.

ADMIXTURE WITH POPULATION SEARCH

Looking for a specific ethnic group? The Admixture with Population Search function (available from the GEDmatch home page) allows me to take a deep dive into a particular ethnicity. For example, if I want to run only DNA tests that test for Ashkenazi populations, I can search for *Ashkenazi* from the Admixture/Oracle Population Search and see all the tests that include those sample populations. I can then click on the project I want and enter my kit number on the bottom, then see how I compare in all of these different Admixture calculators. This is a great shortcut if you are trying to narrow your search.

GEDmatch.Com

Admixture/Oracle Population Search Utility

Select	Project	Calculator	Population
○	MDLP Project	MDLP K23b	Ashkenazi
○	MDLP Project	MDLP K23b	Ashkenazi_Jew
○	MDLP Project	MDLP World	Ashkenazim_V
○	MDLP Project	MDLP World	Ashkenazim
○	MDLP Project	MDLP World-22	Ashkenazim
○	MDLP Project	MDLP World-22	Ashkenazim_V
○	Eurogenes	Eurogenes EUtest V2 K15	Ashkenazi
○	Eurogenes	Eurogenes K13	Ashkenazi
○	Dodecad	Dodecad K12b	Ashkenazi
○	Dodecad	Dodecad K7b	Ashkenazi
○	Dodecad	Dodecad V3	Ashkenazi
○	Dodecad	World9	Ashkenazi
○	HarappaWorld	HarappaWorld	ashkenazi
○	puntDNAL	puntDNAL K10 Ancient	Ashkenazi_Jew
○	puntDNAL	puntDNAL K12 Modern	Ashkenazi_Jew
○	GedrosiaDNA	Ancient Eurasia K6	Jew_Ashkenazi
○	GedrosiaDNA	Gedrosia K3	Ashkenazi_Jew
○	GedrosiaDNA	Near East Neolithic K13	Jew_Ashkenazi
Enter Kit Number:			Continue

With the Admixture/Oracle Population Search, you can find the Admixture tests that have the most relevance to a particular subject, such as *Ashkenazi*.

Toward the bottom of the page, we see a summary by chromosome that looks at all the selected individuals. It displays the number of matching segments, a range, and the largest segment. The higher the number of the total matching segments, the stronger the match. Below that is a comparison of each person with another person (similar to the One-to-One Compare tool): the chromosome where the match is, starting and ending points, the total number of shared segments, and the length of the segment of shared DNA.

Finally on the bottom of the page, you can click on the option to view all that data on a 3D rendering of the chromosome (image **F**). This is a really neat feature, allowing you to look at each chromosome one by one as well as tilt the X and Y axes for a better look.

Multiple Kit Analysis

The powerful Multiple Kit Analysis allows you to compare up to fifty kits and generate a wide variety of data, including both 2D and 3D chromosome browsers, massive matrices comparing amounts of shared DNA, matching GEDCOM files (if you've uploaded any), and lists of matches and matching segments that are vital to DNA triangulation.

The 2D chromosome browser (image **G**) frequently comes up in research. This shows the analysis of matching segments (from start to end) as well as their sizes as compared with each other. "New Root" refers to a newly discovered segment, and the numbers indicate the overlap between the primary kit and the others who are compared to on that root. You may need to verify the matches you find here by comparing both kits to each other separately with a known relative match sharing the same DNA on the same segment. The generational matrix (image **H**) is also interesting, showing both a visual and matrix of the matches. As these are three siblings and a first cousin, the generational estimate to the most recent common ancestor is between 1 and 2.

Personal Tools

Now that we've discussed the free comparison tools offered by GEDmatch, let's look at the remainder of the free analyses. These focus mostly on the test-taker, not potential relationships. While generally less useful than the One-to-Many, One-to-One, and X One-to-One tools, these can still provide interesting and valuable information about your genetic past.

Admixture (Heritage)

GEDmatch has its own ethnicity tool called Admixture that analyzes your upload in its own way to give you yet another ethnicity breakdown, mostly for academic purposes.

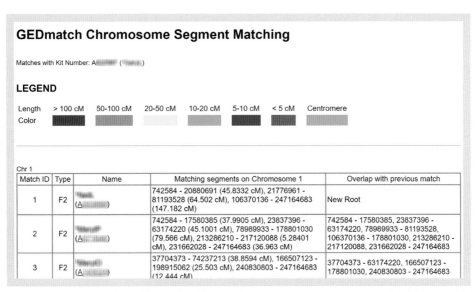

GEDmatch Chromosome Segment Matching

Matches with Kit Number: A████ (█████)

LEGEND

Length	> 100 cM	50-100 cM	20-50 cM	10-20 cM	5-10 cM	< 5 cM	Centromere
Color							

Chr 1

Match ID	Type	Name	Matching segments on Chromosome 1	Overlap with previous match
1	F2	████ (A████████)	742584 - 20880691 (45.8332 cM), 21776961 - 81193528 (64.502 cM), 106370136 - 247164683 (147.182 cM)	New Root
2	F2	████ (A████)	742584 - 17580385 (37.9905 cM), 23837396 - 63174220 (45.1001 cM), 78989933 - 178801030 (79.566 cM), 213286210 - 217120088 (5.28401 cM), 231662028 - 247164683 (36.963 cM)	742584 - 17580385, 23837396 - 63174220, 78989933 - 81193528, 106370136 - 178801030, 213286210 - 217120088, 231662028 - 247164683
3	F2	████ (A████████)	37704373 - 74237213 (38.8594 cM), 166507123 - 198915062 (25.503 cM), 240830803 - 247164683 (12.444 cM)	37704373 - 63174220, 166507123 - 178801030, 240830803 - 247164683

Image G. This compares three different kits. The main kit is my uncle's, and he's compared with his sister, his brother, and his paternal first cousin.

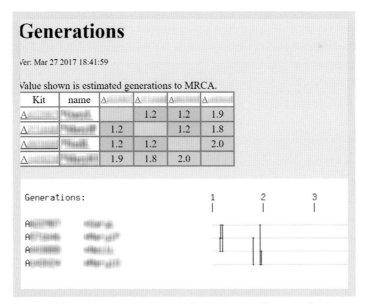

Generations

Ver: Mar 27 2017 18:41:59

Value shown is estimated generations to MRCA.

Kit	name	A████	A████	A████	A████
A████	████		1.2	1.2	1.9
A████	████	1.2		1.2	1.8
A████	████	1.2	1.2		2.0
A████	████	1.9	1.8	2.0	

Generations:

Image H. The Generations matrix shows the kits compared to each other in tabular form, along with a visualization representing the chromosomes.

When you go to this page, however, you are presented with seven different Admixture calculators to choose from:

- **MDLP Project**: This tool provides information about ethnicities from all over the world, similar to those given by the big testing services. Given the broad scope, this option doesn't provide as much detail as some other tools, but it does go into ancient origins to some degree. The number in each test's name indicates the number of populations sampled.

- **Eurogenes**: This Admixture report breaks down European genes. Many of the tests within this category have a number in their names, indicating the number of global populations the test compares your DNA against. The default for those with European ancestry is thirteen populations (K13), but you can select as few as seven and as many as thirty-six. You can also select the Hunter-Gather vs. Farmer test (which examines ancient populations, similar to Family Tree DNA), the Jtest (which specializes in Ashkenazi ancestry), or the EUtest (which looks at all of Europe *except* Ashkenazi Jews).

- **Dodecad**: This Admixture test focuses on Asian and African ancestry, and the various tests examine different combinations of Asian, African, European, and Middle Eastern populations. The World9 test looks at nine worldwide populations, making it somewhat useful even for those without recent African or Asian ancestry.

- **HarappaWorld**: This test is best used by those with South Asian ancestry. It only has one calculator, and it focuses mostly on Sri Lankan, Bangladeshi, Indian, and Pakistan ethnicity.

- **Ethiohelix**: This tool focuses primarily on those with African ancestry. The K10 Africa Only test looks just at ten African populations, while other options allow you to compare nine African populations against a population from another region (namely, France, Japan, and the Middle East/Palestine).

- **puntDNAL**: This test provides information about your ancient DNA from various populations around the world. While not useful for genealogical purposes, this can still provide fun details about where your ancestors were thousands of years ago.

- **GedrosiaDNA**: If you come from Eurasian (European and Asian) descent, these calculators are for you. The various tests cover different portions of the Eurasian continent, as well as both ancient and modern analyses.

Though these ethnicity tests vary in their coverage and utility, they have a few principles in common. In general, you'll want to run tests that analyze fewer populations, as an analysis of many populations generates more speculative results. A targeted approach will yield the most accurate results. You should also use the ethnicity estimates provided by the testing companies to identify what specific ethnic groups your DNA matches up

Jtest Admixture Proportions

This utility uses the Eurogenes Jtest model, created by Davidski (Polako). Questions and comments about this model should be directed to him at his Eurogenes Ancestry Project blog.

Kit Number: T█████ Iteration: 840 Delta-Q: 9.888754e-08 Elapsed Time: 19.18 seconds

Population	Chr--> 1	2	3	4	5	6	7	8	9	10	11	12
SOUTH_BALTIC	8.0	10.6	11.4	2.2	7.2	13.2	7.5	4.9	-	-	3.7	4.4
EAST_EURO	-	-	-	-	-	5.1	4.2	-	12.3	-	15.7	-
NORTH-CENTRAL_EURO	12.7	9.3	2.5	4.4	14.7	2.6	9.4	-	8.9	16.8	3.8	22.8
ATLANTIC	9.2	12.4	2.5	13.0	9.7	8.8	6.6	14.2	-	4.3	4.9	4.3
WEST_MED	4.2	11.7	17.8	13.8	3.6	6.8	1.9	6.1	19.9	6.8	6.5	14.5
ASHKENAZI	21.5	19.4	17.7	20.7	27.9	21.1	34.7	27.3	30.8	38.5	14.5	17.5
EAST_MED	29.0	21.8	16.7	23.7	20.6	34.3	16.4	23.8	15.9	14.3	16.2	19.7
WEST_ASIAN	1.0	2.6	13.4	-	-	2.3	4.4	4.7	-	0.2	13.5	6.5
MIDDLE_EASTERN	6.7	12.1	12.6	18.2	13.9	4.6	9.2	17.5	3.8	12.7	16.6	-
SOUTH_ASIAN	-	-	-	1.3	-	-	1.2	-	1.9	2.3	4.0	2.1
EAST_AFRICAN	1.1	-	4.3	2.1	-	1.0	1.8	1.6	-	-	-	0.4
EAST_ASIAN	0.4	-	1.1	-	-	-	0.6	-	-	-	-	3.6
SIBERIAN	6.4	-	-	0.6	-	-	-	-	4.2	4.1	-	-
WEST_AFRICAN	-	-	-	-	2.4	-	2.0	-	2.2	-	0.6	4.3
Number of SNPs eval:	12770	12474	10821	9302	9763	10278	8698	8684	7867	8642	7802	7970

Image I. My Jtest in the Admixture Proportions by Chromosome analysis shows how I compare with various ethnic groups at each chromosome. This image only shows up to chromosome 12, but the full analysis displays to all twenty-two autosomal chromosomes.

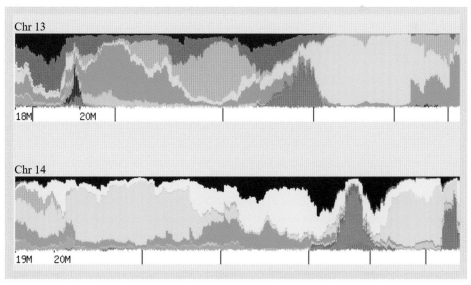

Image J. The chromosome painting, a function of GEDmatch's Admixture, visually displays your ethnic breakdown for each chromosome, with each color representing a different ethnicity of DNA in that chromosome.

with, as the Admixture has tests designed for certain ethnicities. For example, if you are Ashkenazi Jewish, you do not want to run the GedrosiaDNA or Ethiohelix tests.

You might come across the term Oracle in some of these Admixture tests. The concept of the Oracle is used across a number of calculators, helping you narrow down your origins to a specific location or religious group. Oracle's Single Population Sharing looks into a single population that your DNA most accurately matches against and lists the top twenty. You will see how closely you match each group: the smaller the match, the closer the match is. Oracle's Mixed Mode Population looks at two combined populations to see how you match and works under the assumption that you are of mixed heritage. Again, the closer you are toward a population, the more closely you match. The Oracle 4 test is a more enhanced test that combines three or four populations. Look at both Oracle and Oracle 4 to get a better picture, as results may vary across both types of results.

CHROMOSOMAL OPTIONS

In addition to viewing your results through Oracle, you can also view Admixture Proportions by Chromosome (image **I**). This lets you see your results by individual chromosome, plus what percentage of each chromosome is associated with a specific population. Think of it like the 23andMe breakdown per chromosome.

The option below that is Chromosome Painting and Chromosome Painting with Reduced Size. The Chromosome Painting (image **J**) is a "painting" or illustration of your admixture properties, broken out by chromosome and segment. Each chromosome is listed independently, one above the other, and a histogram of sorts illustrates the prevalence of each ethnicity on each chromosome (with each ethnicity indicated by a different color). The Reduced Size option just makes the whole chromosome painting more digestible and easy to view at a glance (image **K**).

COMPARING TWO KITS

In addition to viewing your own ethnicity estimate, you can compare the ethnicity estimates of two kits. This can be useful in helping to identify what ethnicities you and a DNA match share, though (of course) the analysis isn't perfect. And, like with the Chromosome Painting, you also have a Reduced Size option to make the information easier to process. For more information about these categories and the types of outputs you receive, visit <genealogical-musings.blogspot.com/2017/04/finally-gedmatch-admixture-guide.html>. Image **L** shows the comparison between mine and my mother's kits.

Are your parents related?

This tool examines your DNA to determine if there are any overlapping chromosomes between the two sets of autosomes. This outcome, called "Runs of Homozygosity," looks

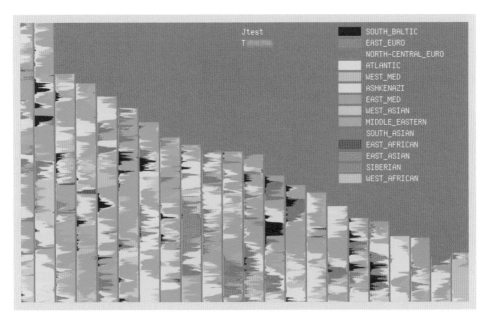

Image K. You can view all twenty-two of your autosomal chromosomes on one display with the Chromosome Painting—Reduced Size.

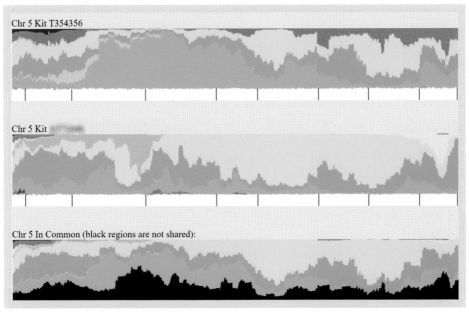

Image L. Though Admixture is mostly just to analyze one person's DNA kit, you can compare two kits and display them in a chromosome painting. My kit is at the top of this image, my mother's is in the middle, and the combined result is on bottom. The black indicates DNA we don't share.

Haplogroup		Autosomal			X-DNA	
Mt	Y	Total cM	Largest	Gen	Total cM	Largest
U4a3a		3587.1	281.5	1.00	196.0	196.0
	J1a3a	3587.1	281.5	1.00	196.0	196.0
U4a3a		3587.1	281.5	1.00	196.0	196.0
N1b1b1	R-Y19847	3586.9	281.5	1.00	196.0	196.0
U4a3a		3586.9	281.5	1.00	196.0	196.0
U4a3a		1961.8	131.2	1.44	28.9	23.2
U4a3a		1897.1	113.9	1.46	30.5	24.9
N1b1b1		1869.9	218.6	1.50	196.0	196.0
	R-Y19847	1735.9	214.5	1.50	0.0	0.0

Image M. The One-to-Many Tier 1 tool has a much cleaner interface than its free counterpart. This screenshot zooms in on the haplogroup, autosomal, and X-DNA columns. Each row represents a different kit, and the first columns (not shown) provide the owner's kit number, username, and e-mail address.

at large SNP blocks and sees if there are matching alleles across the same SNPs. To put it another way: It examines your DNA to see if any of it overlapped before you were born.

For example, I know my husband's great-grandmother and great-grandfather were first cousins. When I look at this analysis for my husband's great-uncle (the son of these two first cousins), he's flagged as having ancestors who had a genetic relationship with each other. The tool looks at each chromosome and indicates the cM shared across each individual chromosome, then displays the largest segment, total number of segments, and estimated number of generations to the most recent common ancestor. In my husband's great-uncle's case, he has a 197.1-cM match that overlaps across thirteen segments. Compare that with my paternal grandfather, who has a single segment overlap of 9.2 cM. Given that low amount, we're unlikely to discover how his mother and father are related.

This is another analysis tool that doesn't necessarily have much utility, but can still be an interesting piece of data.

GEDmatch Tier 1 Tools

All the features we've discussed thus far are free, but GEDmatch also offers Tier 1 functionality for a low monthly (or annual) cost. These advanced tools perform more sophisticated tasks, and so require a small fee to access. You'll likely find (like I did) that the tests are well worth the cost.

Let's take a look at each to see what they can do.

One-to-Many Matches (Tier 1)

The Tier 1 One-to-Many Matches feature is a more advanced version of the One-to-Many tool that sports a different interface, one that feels less cluttered and uses an easier, more readable sans serif font. As you can see in my results (image **M**), this tool provides identical information to the free One-to-Many but in a different format. I also can choose to see more than the default two thousand results—up to one hundred thousand. That said, it's not a drastic change from the other interface. I can also sort by name, e-mail address, recency of matches, and more. And, like the free One-to-Many chart, my results aren't absolute—I should still check each match using the One-to-One tool to confirm the strength of the relationship.

The first text column shows the matching kit number, followed by user's name and e-mail. The next column shows a link to an uploaded GEDCOM, if provided. This view also provides the age of the kit (i.e., how many days since the kit was uploaded to GEDmatch).

The rest of the data mirror that in the free One-to-Many analysis: the chip number used for the upload, the gender of the person whose file was uploaded, the mtDNA and Y-DNA haplogroups (if entered), autosomal data (total cM, largest cM segment, and the approximate generation to the most recent common ancestor to another tenth of a degree), and X-DNA data (total cM and largest cM segment).

You can also select several kits, then click Visualization Options to access the Multiple Kit Analysis feature we've explored earlier in this chapter. The Visualization link points you back to the free tool.

Matching Segment Search

Here, you can enter a kit number and discover who in the GEDmatch database matches it. In other words, this tool allows you to see anyone's DNA matches compared to a particular kit. Simply enter the kit number, set the threshold, and go.

The Matching Segment Search (image **N**) looks at each chromosome and identifies segment matches that overlap, allowing you to find common relatives. This analysis shows the start position and end position, the number of cM in this 20-plus cM segment, the number of SNPs used for the comparison, the name of the match, the gender, and the other user's e-mail address. Each chromosome segment is color coded, and the list goes from the first part of the segment to the end of the segment, and then on to the next chromosome.

This tool takes some time to run (it has to process a tremendous amount of data), but it's useful in comparing any known matches with overlapping segments so you can work together to find a most recent common ancestor.

GEDmatch
Matching Segment Search - V2.1.2

Using nfs kits

Kit: A▓▓▓▓ (F2) (▓▓▓▓)

Minimum threshold size to be included in total = 700 SNPs
Minimum segment cM to be included in total = 20.0 cM
Please wait for matching kit search. This may take a few minutes...
Looking...
Matching segments will be identified between you and your closest 9976 matching kits
Segment comparison analysis progress is shown by a string of asterisks ('*') on the lines below. Each asterisk is 10 comparisons.

```
**************************************************************************************
**************************************************************************************
**************************************************************************************
**************************************************************************************
**************************************************************************************
**************************************************************************************
**************************************************************************************
**************************************************************************************
```

Kit	Chr	Start Position	End Position	cM	SNPs	Name	Sex	Email	Segments
A▓	1	742,584	34,769,000	62.8	6,240	▓▓	F	▓▓	
A▓	1	742,584	17,747,430	38.4	3,320	▓▓	M	▓▓	
T▓	1	10,737,525	31,112,157	35.5	3,442	▓▓	M	▓▓	
H▓	1	35,033,356	56,570,342	21.8	2,725	▓▓	U	▓▓	
H▓	1	36,846,938	60,554,874	26.5	3,168	▓▓	F	▓▓	
T▓	1	37,182,595	59,419,566	24.9	3,178	▓▓	F	▓▓	
H▓	1	59,262,391	85,790,073	26.1	4,030	▓▓	M	▓▓	
T▓	1	59,457,853	152,048,512	65.1	9,692	▓▓	M	▓▓	
H▓	1	77,644,730	114,260,718	35.2	5,570	▓▓	F	▓▓	
M▓	1	110,101,214	161,801,809	32.8	3,825	▓▓	F	▓▓	
T▓	1	110,102,783	161,744,911	32.6	3,825	▓▓	M	▓▓	
T▓	1	151,512,249	166,751,129	24.1	2,933	▓▓	F	▓▓	
A▓	1	154,934,023	182,035,908	34.9	4,505	▓▓	F	▓▓	

Image N. GEDmatch's aptly named Matching Segment Search identifies potential chromosome segments that your kit matches.

Relationship Tree Projection

This experimental tool calculates possible relationships of two kits by looking at genetic distances between autosomal and X-DNA. To use the tool, you need two kits that share X-DNA. From the Relationship Tree Projection link, enter the two kit numbers (in this case, mine and my third cousin once removed), each test-taker's sex, and the total shared (autosomal) cM, largest segment of (autosomal) DNA, the total cM of X-DNA, and the largest segment of X-DNA. You'll also specify whether one donor is older or younger than the other. You'll likely want to leave the rest of the settings the same.

Your results will be a text summary of the data you've provided, in addition to calculated values and assumptions. The output includes the estimated number of generations from donor 1 and donor 2 to the common ancestor as well as the proposed genetic relationship. In one case, the relationship was estimated as a second cousin once removed—but in reality, we know he is a third cousin once removed. Therefore, don't take the estimated relationships at face value, but assume they're close.

Your results also come back as a "family tree" that shows what your actual pedigree may look like, based on the amount of shared information. Image **O** shows a snippet of that tree, which is too large to display on my computer screen, let alone in a book. Notably, the tool will report the estimated number of generations from each kit (Donor) to the most recent common ancestor, plus a proposed cousin relationship between the two.

What about the data in each individual box? Don't worry about the I or J values—these values are debugging data from an early version of the program. The A and X values, however, do mean something—they represent the maximum cM values for autosomal DNA and X-DNA, respectively. These are a crude way of understanding familial relationships between the two selected kits based on DNA inheritance patterns. Every generation reduces the A value by 50 percent, and the X values are reduced based on X inheritance, which are determined by the gender of the individuals involved in that particular inheritance pattern for each generation. Group numbers are assigned within the tree branches as the algorithm considers possibilities. Below this tree are proposed relationship paths based on the different groups found in the tree (but not necessarily all of them) that show the estimated male and female autosomal and X-DNA most recent common ancestors.

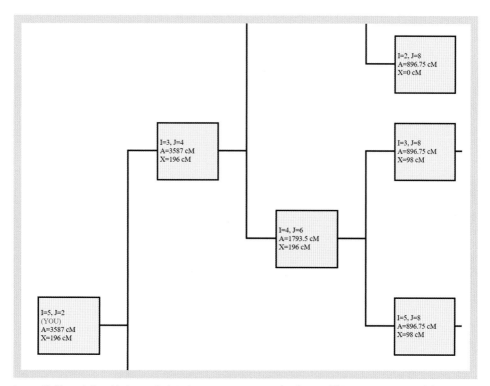

Image O. The relationship tree calculator's output generates a family tree. This is just a section of the tree projection for my DNA kit—the full image wouldn't fit on this page.

Lazarus

Lazarus is a unique tool that allows users to "create" a close ancestor out of available DNA relatives (cousins, siblings, aunts, uncles, etc.). Lazarus helps create surrogate kits for deceased or missing relative by reconstructing part of his DNA (a "resurrection" of sorts). For example, my maternal grandparents are no longer alive, but I can put together a picture of their DNA by testing their living relatives. Lazarus may only be able to provide you with some of the genome (not all SNPs), but more than enough to be helpful in figuring out the general direction of the relationship. However, as GEDmatch's own wiki acknowledges, it's not unheard of for you to use over one hundred kits to produce a single Lazarus kit for a more complete picture of the genome.

My first objective is to re-create my grandmother's genome. I can begin by naming the kit and selecting a segment threshold (both in SNPs and cM), and the sex of the person whose kit I'm creating (image **P**). Then, I need to fill out the two sections on the bottom of the Lazarus landing page, beginning with direct descendants, then moving on to siblings, parents, and cousins. Besides naming her kit and adding kit numbers, I left everything else at their defaults.

Because I only have four known relatives of my grandmother, I was only able to derive partial information. I am hoping in the near future to test my grandmother's maternal side (I only have data on her father's side to date) to get more accuracy. Still, I can use this kit to search for DNA matches. On the left side of the GEDmatch home screen, when you click on Edit or Delete your DNA resource files, you will see the letter R next to these kits. This identifies them as research kits and are strictly used for you to research matches privately. Note that Lazarus-created kits (like phased kits) will never show up in One-to-Many matches, but they are available for searching privately.

RESEARCH TIP

Upload GEDCOMs to GEDmatch

Did you know you can upload GEDCOMs to GEDmatch? Despite the similarity in names, this feature is not overly utilized on GEDmatch—but it should be. With GEDCOM files, GEDmatch users can compare their family trees to yours to see what relationships you may have in common. You can access GEDCOMs under the Genealogy—Family Trees and Genealogy tabs on the home page. Start with the GEDCOM Genealogy Upload link. Once you've uploaded a GEDCOM, you can compare it to others in GEDmatch's database (under 1 GEDCOM to all) or to another specific GEDCOM (under 2 GEDCOMs). You can also search all GEDCOMs under the appropriate link or see what GEDCOMs match your DNA results under GEDCOM + DNA matches. If you choose to upload a GEDCOM, it will also be linked on your One-to-Many results.

Image P. The Lazarus tool allows you to essentially reconstruct a DNA kit by providing information about relatives.

Triangulation

We've referred to triangulation so much in this book, but we haven't really gotten into the nitty-gritty of what it is. GEDmatch has a Tier 1 Triangulation tool that may help you get started. We'll still explore triangulation methods in the next chapter.

First of all: What is triangulation? In its simplest form, triangulation is a technique where you identify one or more DNA segments that are in common among three individuals. In this context, we could find two people who are listed as DNA matches, plus a third who matches with both individuals on the same segment. Assuming you know the most recent common ancestor for the two of them, you can reasonably assume the third one comes from that line and pursue that line of research.

For example, Jerry and Tonya are second cousins (their grandparents were siblings) and share a big match on chromosome 12. Herbert also overlaps (albeit with a smaller segment match) with Jerry on the same chromosome. If Herbert and Tonya also overlap in the same area on chromosome 12, we can deduce that the relationship is coming from the set of great-grandparents shared by Jerry and Tonya—somewhere up the line, at least. If third cousins also compare and overlap with Herbert, Jerry, and Tonya on that same segment, then we even know which great-grandparent it is. Remember, though, you need to compare each test-taker to the other as well as both of them to each other. In the case of only Herbert, Jerry, and Tonya, we'll need to do a One-to-One match with Herbert and Jerry, Jerry and Tonya, and Herbert and Tonya. If they all share that segment, that's a triangulated match.

Be careful with triangulation, as you don't want to prematurely declare two people a match. (Additionally, you should never do triangulation on the X chromosome, because X-DNA doesn't act like autosomes.) Three people sharing DNA on the same segment aren't necessarily all matches with each other. For example, Sally doesn't match John just because John matches Scott and Scott matches Sally. John also has to match Sally—in the same exact place on the same chromosome, as in the example with Herbert, Jerry, and Tonya. You'll need to do all of those comparisons and analyze other data and relationships to confirm or deny that John and Sally match each other in addition to Scott. If you confirm that match and they have a good tree up into the further generations, you can put two and two together to build your family tree further.

Now what about the GEDmatch Triangulation tool? This looks at your top matches within a certain threshold and compares matches against each other. GEDmatch then

Chr	Kit1	Kit2	Start	End	cM
1					
1	A	A	15,235,810	34,287,917	28.1
1	T	M	17,765,403	24,430,347	12.4
1	A	T	19,859,881	28,085,744	9.8
1	A	T	19,859,881	28,085,744	9.8
1	A	T	21,856,198	27,869,722	7.3
1	A	A	36,364,199	161,723,972	106.2
1	A	A	36,718,382	45,594,065	11.1
1	A	A	36,718,382	45,594,065	11.1
1	A	A	38,719,726	56,098,881	15.7
1	A	M	39,125,091	53,819,144	11.4
1	A	M	39,125,091	53,819,144	11.4
1	A	A	43,044,833	56,098,881	10.4
1	A	A	43,044,833	57,897,708	13.3
1	A	A	43,902,366	119,719,096	74.0
1	A	A	43,902,366	119,719,096	74.0
1	A	A	44,362,168	97,413,151	51.1
1	A	A	44,362,168	97,413,151	51.1
1	A	A	54,789,591	97,413,151	42.2

Triangulated results sorted by Chromosome, Start Position

Image Q. With your triangulation results, you'll view kits that GEDmatch has targeted as being eligible for triangulation research.

displays results by chromosome and position, kit number, and in tabbed or graphical format. Be patient and don't close the browser when this is running. The analysis can take between thirty and forty-five minutes, as GEDmatch has a lot of data to crawl through!

When the results are finally available, you will see a chart that looks like image **Q**. Here, you can see which kit numbers GEDmatch has identified as belonging in triangulation groups with the specified kit number, along with the amount of shared DNA and the start and end points. You can save this data to Excel, but (like the processing itself) it will take GEDmatch a while to get all of that data out of your clipboard and paste to your spreadsheet tool.

Now that you have your data, what do you do with it? Thinking about the earlier examples, let's assume you have three people who overlap with you on chromosome 4. You will need to do several (in this case, seven) One-to-One comparisons to validate the triangulated relationship: you to person 2, you to person 3, person 2 to person 3, etc. If there's overlap on one segment on the same chromosome in all of these tests, you can confirm you're all related to each other. You can reach out to these potential matches and see if you can find a common ancestor by working backward. That will give you a line from which to dig further and find your birth family. I'd recommend contacting only matches with larger segments (e.g., 10 to 15 cM).

You never know—if you reach out to two different people with overlapping segments, maybe those two know each other already. The knowledge they share can point you directly to other relatives, potentially saving you a lot of time. By collaborating, you may find the missing connection.

Triangulation Groups

The Triangulation Groups tool allows you to take the next step in your genetic research, identifying groups of other test-takers who you should analyze for DNA triangulation. As with the Triangulation tool, you'll need your kit number, the number of kits you want read (more will be slower), the minimum segment size, and the upper limit of related matches to use, where the higher values indicate close relatives (as per the charts in chapter 4). Given how sprawling this analysis can be, you might consider raising the threshold. Like the Triangulation tool, Triangulation Groups also takes a long time to process depending on the number of kits you are reading for the comparison.

After processing, you will receive three key data visualizations: an estimated tree for each triangulation group, a bar that breaks down each chromosome, and a CSV file you can download to further analyze your results.

Image **R** shows triangulation groups results from my grandmother's kit (top left, under reference). Each bar graph in the left-hand column represents how my grandmother compares to another user on all segments of autosomal DNA, from chromosome 1 to

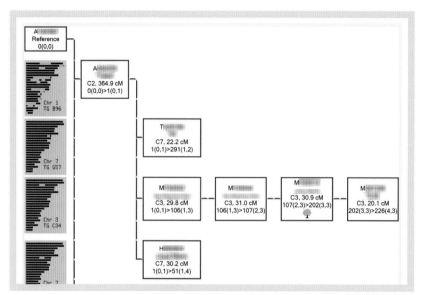

Image R. The Triangulation Groups output organizes potential triangulation DNA kits by chromosome.

chromosome 22. (Remember, we don't look at X-DNA when triangulating.) The first chart shows how her DNA compares to one other test-taker (her paternal first cousin), with the green indicating overlap and black indicating no overlap. This first chart is heavily green, since I'm comparing her DNA to that of her first cousin (a close relative).

Each red box indicates a different DNA kit that has been compared to my grandmother's. Within this box is the kit number, the person's name or alias, the chromosome number where the DNA match is (e.g., C3 for chromosome 3), and the total size of all segments (in cM) that triangulate. (You can ignore the row with the numbers in parenthesis. It's leftover code from debugging.) If you hover over the box, you can also view the start and end points for the segment match (and others, if they exist). A tree indicates that the user uploaded a GEDCOM or WikiTree file.

The boxes are organized to represent each kit's genetic relationship with the others. Blue lines connect boxes that are related, and the red boxes line up with the bar graph triangulation group they belong to.

My grandmother's kit in the image happens to be connected to three ancestral lines or triangulation groups: One stems from the C2, 364.9 cM box to the right of the chromosome 1 bar chart (her first cousin), and the two others are on chromosome 5 and chromosome 7 (not pictured).

Let's take a look at that first one: The triangulation group including my grandmother and her first cousin. Each of the red boxes connected by a blue line are in a triangulation group with my grandmother and her cousin, though (indicated by their placement) they match in different ways. In the third column, we see three groups of DNA kits that share DNA on various chromosomes: the first matches on chromosome 7 with a 22.2 cM segment; the second contains four individuals who all match my grandmother, her paternal first cousin, and each other on chromosome 3; and a third isolated branch that consists of another individual who overlaps with my grandmother and her cousin on chromosome 7.

All of these individuals who are nested under my grandmother's first cousin are triangulated matches to my grandmother and her first cousin: They match on the same segment in the same place, thereby indicating all six have the same common ancestor. You can view which segments each test-taker shares with my grandmother and her cousin by consulting the corresponding bar chart in the first column. I ran a One-to-One comparing one of the nested kits to my grandmother's and the amount of shared cM holds up (though you might some variation since the tests use different algorithms).

However, other branches that don't connect via a blue line through that first cousin are related in different ways. Those relatives (who would branch from a different test-taker in the second column) would come from a different shared ancestor.

You can also view your data in graphic bar format. Here, you can view much of the same data as previously mentioned, including the start (From) and end (To) points, length of the largest segment of shared DNA (cM) and the number of shared SNPs. The table is organized by chromosome, and the green bar indicates the length and placement of the shared DNA. This is also where you can more easily visualize overlap.

Image **S** shows my grandmother's results to matches on chromosome 7. The long green line (in the row denoted B96) represents my grandmother's first cousin. We also see two matches (denoted as G36 and G57) that fall under my grandmother (the person this whole chart is compared to) and her paternal first cousin (as B96, listed under the left-most From column). These are two separate triangulation groups, as the person in group G36 doesn't overlap with the person in group G57. As they are smaller but significant in size, we likely have one or two common ancestor(s) up the line, each that matches my grandmother and her paternal first cousin. Now, it's up to me (or one of the others) to figure out if we can find that connection.

Chromosome 7										
From	TG	Kit	Name	email	Chr	From	To	cM	SNPs	
	B96	A▓▓▓▓ ▓▓▓		▓▓▓▓▓▓	7	33438655	152658585	24.6	47896	
B96	G36	H▓▓▓▓ ▓▓▓		▓▓▓▓▓▓	7	81239365	115583490	30.2	14491	
B96	G57	T▓▓▓ ▓		▓▓▓▓▓	7	136570463	150623159	22.2	6271	

Image S. You can also view your triangulation groups as a bar chart.

We'll go into more detail on how to use this data in chapter 11. But for now, just know that these tools can help you better identify potential genetic matches.

My Evil Twin Phasing

The last Tier 1 tool currently available is called "My Evil Twin Phasing." It takes inputs from a child and at least one of his parents to create a phased kit representing the DNA that the child did not inherit (ergo, the child's "evil twin"). Obviously, you'll need all three DNA kits to run this analysis. Simply enter the appropriate kit numbers. GEDmatch will then generate a constructed kit representing the "evil twin" of that child, and you can then use that for One-to-One or One-to-Many tests if desired. Run this analysis if you aren't finding genetic relatives for a person and suspect it's because he didn't inherit a traceable quantity of a certain piece of DNA. The phased kit is essentially an artificial sibling that you can use to identify additional genetic matches.

11

Triangulating Your DNA Data

No book on adoption genealogy would be complete without a chapter on triangulation—that is, looking for one or more DNA segments that at least three people share. With that information, we can conclude that they share a common ancestor. We scratched the surface of triangulation in chapter 10, when we discussed the triangulation tools offered by GEDmatch **<www.gedmatch.com>**. With these tools, we could group together people who match DNA on the same segment on the same chromosome. This often indicates the existence of a common ancestor.

But we'll examine triangulation in much more detail in this chapter, taking you through the process and looking at tools to help you find genetic relatives. For the purposes of our studies, we are going to focus on autosomal triangulation, though there are applications of triangulation for Y-chromosomal DNA (Y-DNA) to find out the real haplogroup of a distant ancestor.

The Basics of Triangulation

To do triangulation properly, you will need to map out your chromosomes and compare them to another test-taker. Then, you look at a third person who overlaps with you and that other test-taker. You'll need a chromosome browser to view the overlap between three individuals, or you could use GEDmatch (which, in my view, is the better option). Note that triangulation does not work with full siblings, but you can triangulate with half siblings if you know whether they're maternal or paternal. Despite the name, you don't have to limit triangulation to three people. A fourth or fifth person can "triangulate" to the others in the group.

Search Both Sides of Your Family Tree

Though triangulation can help you find genetic matches, you may need to do extra work to determine if a match is related to you via your mother or via your father. Sometimes you will need to use two other potential genetic relatives—known cousins on the other parent's side—to tell you whether a third match is maternal or paternal.

Triangulation works by looking at your DNA that matches the DNA of other test-takers on a particular segment, then determining if all individuals match each other on that same spot. If all do match, you can work to identify the most recent common ancestor (MRCA) who has contributed that segment to all in the triangulation group. If they don't, you can assume the match shares a different ancestor. You can also use maternal versus paternal triangulation groups to help you determine which side of your family a match comes from: A test-taker is paternal if he doesn't match with your maternal triangulated group, or he's maternal if he doesn't match with your paternal triangulated group.

Long paper trails make triangulation easier, and (in general) triangulation gets more difficult as the overlapping segments become smaller and smaller. Testing more-distant cousins can also help make your research easier. Older cousins, in particular, can add valuable data to your research, as they share more DNA with ancestors than relatives from more-recent generations.

Since GEDmatch is totally free, it's a great place to start your triangulation process (assuming, of course, that your matches have uploaded their data there). The One-to-One comparison on GEDmatch allows you to properly visualize your raw data for triangulation. For example, let's say you run a One-to-One comparison between you and Bob, you and Jane, then Bob and Jane. GEDmatch reports you match Bob on chromosome 1 between 6,400,121 and 70,232,333, and that Bob matches Jane on chromosome 1 between 4,989,198 and 65,991,843. If you *also* match Jane on chromosome 1 between start point 7,333,875 and 44,124,520, you can see a definite overlap. The segment that you, Bob, and Jane all share can help you identify the ancestor who you three have in common. See chapter 10 for more on uploading your raw DNA data to GEDmatch and using the site's tools to analyze it.

Until the creation of some great chromosome mapping tools, my preferred method of charting was manually using GEDmatch or a chromosome browser. In a spreadsheet (I use Google Docs to share with others and collaborate), I name the tab after the primary person who all matches are being compared to. In separate columns, for every relative I test, I list the match's name, the type of match (M for maternal, P for paternal, and blank for unknown), the chromosome the match is on, the start and end points, the size of the match in centimorgans (cM), the total number of single nucleotide polymorphisms (SNPs) used for the comparison, the MRCA(s) I know, and comments. These notes can be anything from the kit number used in the comparison to info on overlapping segments with another cousin that prove the match is either maternal or paternal.

I freeze the top, header row (allowing me to see the header no matter how far I scroll down), and I highlight any overlapping segments with two or more people since these form a triangulated group. (Remember, you will need to compare everyone against each other—seven comparisons for four people, for example.) Those who match you on a single chromosome in the same spot are potentially members of a triangulation group who you'll want to study. Generally, you'll only want to consider matches who share large segments of more than 15 cM (or 20 to 25 cM, if you're from an endogamous culture). This entire process (though time-consuming) works surprisingly well, especially when processing new potential DNA matches.

You can use 23andMe **<www.23andme.com>**, MyHeritage DNA **<www.myheritage.com/dna>**, and Family Tree DNA **<www.familytreedna.com>** to help you identify individuals with their chromosome browsers, the ICW tools, and the matrix tools (if available). But you'll have trouble comparing data from across different testing services, particularly if you don't have the appropriate relatives on an individual service. (AncestryDNA **<dna.ancestry.com>** doesn't provide a chromosome browser, so that service's analysis tools won't be very helpful for triangulation. The closest thing it has to triangulation is its Circles feature, which is contingent upon the ancestor being entered into each user's tree.) I highly recommend that anyone who tests on any of these services uploads to GEDmatch, because having data in one central location helps validate hypotheses for both you and others looking for relatives.

One quick note: The In Common With (ICW) tools offered by DNA testing companies do not provide data of the same relevance and accuracy as triangulation. In fact, they should never be used for triangulation. The matches you and your matches share in common don't necessarily share DNA (and thus, relate to all of you). In other words, ICW lists include people who have matches in common with each other, but they may not overlap at all and thus cannot be considered for triangulation. Instead, you need to rely on the overlapping segments of DNA.

Case in point: A distant cousin reached out to find her adopted mother's parents. Her closest matches were close cousins of mine. When looking closely at the data, the distant cousin matched significantly with cousin #1 and also significantly with cousin #2, but the segments she shared with cousin #1's were not the same segments she shared with cousin #2. Despite being connected by DNA, these matches did not triangulate.

Tools for Triangulation

Now that we've discussed the broad strokes of DNA triangulation, let's dig into the nitty-gritty: what tools you can use to triangulate DNA matches, and how you can use them.

DNA Painter

DNA Painter **<dnapainter.com>** comes highly recommended in helping users map out their ancestry with its own unique spin on a chromosome browser. The tool stores your data directly in an online database that helps you chart out locations on your chromosomes and associate them with shared ancestors. Its creator, Jonny Perl, was awarded the top prize at the 2018 Innovation Contest, part of the RootsTech genealogy conference.

The site has a number of self-explanatory introductory videos to help you navigate the service, but using the site is simple. All you need to do is paste in match data from a chromosome browser or GEDmatch's One-to-One tool (image **A**). In this image, I am pasting GEDmatch One-to-One data to chart information about my grandmother. All you need to do is highlight your results (the entire chart in the One-to-One comparison, with the top and bottom lines optional) and paste to the window that pops up when you click on "Paint a Match."

The shaded area behind this box represents different chromosome matches that I can associate with known ancestors, but I can also put down unknown ancestors. These will assist me with close relatives whose relationships I can't map out just yet. Eventually, the goal is that someone will triangulate with them, and you'll be able to ascertain the connection to the ancestor the three of you share.

To paste data from 23andMe, simply grab the data from the DNA Relatives feature **<you.23andme.com/tools/relatives/dna>** by selecting yourself and the relative you want to know about. (You'll only be able to do this successfully if you are sharing your DNA results.) Right below that, you will see detailed segment data. Simply copy that data as seen on the page and paste it exactly into the DNA Painter screen for your match. In a similar vein, add data from Family Tree DNA by selecting the match you want to explore

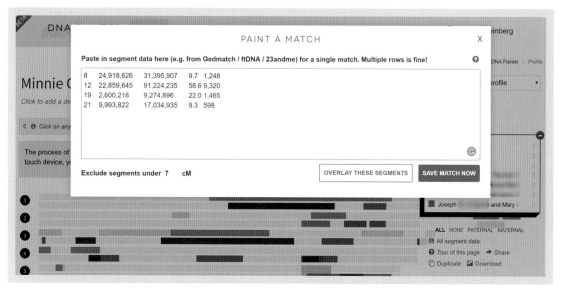

Image A. DNA Painter accepts GEDmatch output straight out of the box to create an image of your chromosome matches.

Image B. DNA Painter, a chromosome mapper, allows you to add labels to individual segments of DNA, helping you visually tie segments to a recent common ancestor.

further, then opening the Chromosome Browser. Click on "View this information in a table," and you'll get the output you need for DNA Painter. Finally, on MyHeritage, when you have your matches open, click on Review Match, then find the chromosome browser. From the Advanced Options menu, download the match data. Then just paste that into DNA Painter, and you're good to go.

The beauty of DNA Painter is that it allows you to use data from all different sources to visualize how you may match someone, plus whether the match is maternal or paternal. You can link ("paint") particular familial branches to matches, such as a surname, "Maternal," "Paternal," or "Unknown," plus add notes. When you hover over any segment that you've painted onto the chromosome, you can learn which ancestor it comes from (image **B**). As you can see, I have a few matches in this profile who are undetermined—I am not sure if they're maternal or paternal, so I'm marking them as "Unknown" until I can find a closer match.

DNAGedcom

The DNAGedcom **<www.dnagedcom.com>** desktop app, available for both PCs and Macs, lets you download your data from a number of websites to access additional analysis tools. Depending on which testing service you're importing data from, DNAGedcom can perform different analyses:

- 23andMe: Collect matches and ICW matches that have shared segments or match a specific individual
- AncestryDNA: Collect matches, trees, and ICW matches
- Family Tree DNA: Collect matches, ICW matches, and trees from all who have uploaded their GEDCOMs

Once you download your data (which can take hours, depending on how many searches you have), DNAGedcom will produce files that you can't actually view to understand. Rather, DNAGedcom results merely format your data so it can be used in other applications and tools, including those on DNAGedcom. Let's look at three in particular: the Autosomal DNA Segment Analyzer, JWorks, and KWorks.

AUTOSOMAL DNA SEGMENT ANALYZER (ADSA)

As the name implies, ADSA analyzes your GEDmatch or Family Tree DNA data and creates tables with overlapping segments. This helps consolidate your triangulation research, as you can view everything on one spreadsheet rather than having different documents scattered throughout your computer.

ADSA takes Family Tree DNA data as is, but you'll need to format your GEDmatch data in a particular way to access the tool. To do so, you must be a GEDmatch Tier 1 member (see chapter 10). From your GEDmatch account, run a Matching Segment Search (under Tier 1 utilities) and choose No under Show Graphic Bar for Chromosome. This expedites the process, as well as makes the output easier for DNAGedcom to process. Simply copy and paste the entire Matching Segment Match data (including the chart) into the white box of the DNAGedcom ADSA page, then hit load. Repeat this task with GEDmatch's Tier 1 Triangulation tool. Enter your kit number; keep the default of 3,000 but choose the section option (show results sorted by kit_number, chromosome, segment start position). Push the Triangulate button and wait for the results to process. Next, return to DNAGedcom and click Clear. When the results finish loading on GEDmatch, copy the entire page (the text and the chart), paste it to the DNAGedcom box, and click Load. After all data is locked and loaded, click Download Kit. Now return to the top menu under Autosomal Tools to access the ADSA program.

Once you've uploaded your kit, simply select it from a dropdown and run your report. You can access a number of features within this tool:

- **Classic ADSA**: This runs a report showing segments sorted by each chromosome's number, start points, and end points.

- **Matches**: This produces a big table of match information that allows you to reorder rows, include relative groups (close, cousins, remote), and highlight rows with one or more surnames that appear in the Ancestral Surnames field or rows that have been added since a certain date. If you're from an endogamous culture, you should uncheck the relatives option; if I keep all three options checked, I have 11,342 matches, but that drops to a more manageable 1,915 matches if I leave the relatives option unchecked.

- **Segments**: A variation of the Classic ADSA, this tool features additional input fields to add/remove close relatives, cousins, or remote matches (again, a must-have if you are in an endogamous culture), and highlight matches by surname or date.

- **Expert Mode**: This highly technical output is truly for experts, and the details of it are beyond the scope of this book. If you feel you've mastered the basics of DNAGedcom and want to dive deeper, learn more at **<dnagedcom.com/adsa/ adsamanual.html.php#expert>**.

- **Database Utility**: Here, you can get information on all kits in your database, which include row counts for each kit's list of matches, list of segments, and list of ICW matches. You can use the Database Utility to troubleshoot failed download attempts, such as when use see zero for any matches.

Once you run the ADSA, you can view your results in a number of ways. The default graphs show all twenty-two autosomal chromosomes, plus the X chromosome. You can also view a single chromosome and even zoom by specifying your desired start and end points. You can also manage the display by selecting the "width of segment graph in pixels." Wider pixels (px) allow you to see the segments more easily. However, be mindful that if you print out the results, your results will be cut off. The default of 500px is probably fine, but if you have too many segments shared on one chromosome, you may need to reduce the size. In that case, it may be better to print results in landscape mode (as opposed to portrait mode) or even save your results as a PDF.

You also have the ability to display raw data in a table, so it can easily be copied to a spreadsheet to create a segment data and ICW matrix. Without checking this box, the output is graphical and incompatible with Excel/any spreadsheet document. Just be mindful that the results you get won't be as pretty; they are, after all, intended for processing in a spreadsheet, and that's never aesthetically pleasing.

ADSA uses a segment minimum of 10 cM both for processing purposes and to filter out the irrelevant results. As more and more people use DNAGedcom to upload Family Tree DNA and GEDmatch data, that threshold may increase even more. Within the tool, you can also specify the minimum number of SNPs to use in a segment—more is usually better, but it defaults to 500.

If you're uploading Family Tree DNA results into the ADSA, you can also access a few additional features.

You can run the analysis to include only certain relatives (e.g., close, cousins, or remote) or sort your data by name, shared cM descending, and data descending. You can also highlight matches with Ancestral Surnames by inserting the appropriate surnames (or several, separated by commas). If you're luckier, you may even have matches who provide locations with their surnames. Note that, as on Family Tree DNA, the surnames will encompass any part of the name entered. For example, if you are searching the name Morris, results like Morriston or Morrison will show up in the results. On the other hand, if you're narrowing down to a suffix, like *-ton*, you will not get results like Worthington, Holton, Huntington, or Washington—it only uses the beginning of the word.

Assuming you use ADSA regularly, you may also want to highlight matches that show after a specific date, as your results will continue to change as the database grows.

Let's take a look at some actual results so we know what to look for. If you run Classic ADSA or the Segments reports, you will get a report that looks like image **C**. I've selected this example to analyze just chromosome 1. Each row refers to a specific kit, and the various columns denote details: the name of your match, the start and end points, the cM size, the number of SNPs used for the comparison, and an e-mail address. The center of the table presents an ICW matrix, plus a graphic representation of matching segments spaced along the chromosome from left to right.

CHROMOSOME 1
78 matching segments
Longest is 188.50 cM, Graph = 492 KBP/pixel

KIT	MATCH NAME	START	END	cM	SNPS	EMAIL	ICW
A		742584	10933589	23.50	1902	@gmail.com	
A		742584	15176506	33.90	2643	@gmail.com	
A		742584	17530121	37.90	2985	@gmail.com	
A		742584	18450662	41.00	3253	@gmail.com	
A		10771286	17938230	15.90	1220	@gmail.com	
A		10847297	18069387	16.20	1215	@gmail.com	
A		17765403	214301697	188.50	23380	@gmail.com	
A		29347928	40877062	17.00	1717	@gmail.com	
A		34027288	49819328	16.80	3450	@aol.com	
M		34363656	53316877	17.80	3725	@gmail.com	
A		34578891	60145546	28.60	3322	@gmail.com	
A		34707622	176294306	124.70	15780	@gmail.com	
M		35042482	55160225	19.90	4111	@gmail.co...	
T		36846938	60263046	26.30	5447	@	
M		37182595	59419566	24.90	5091	@	

Image C. The ADSA showcases an ICW matrix comparing your DNA matches to those of several matches in the DNAGedcom database.

DATE 10/07/2017
10 matches

MATCH NAME	EMAIL	DATE	SHARED	BLOCK	RANGE	RELATIONSHIP	KNOWN
		10/07/2017	78.15	14.69	3rd Cousin - 5th Cousin	4th Cousin	14
			16.77	16.77		Cousin	13
						Cousin	14
						Cousin	14
						Cousin	14
						Cousin	14
						Cousin	14
						Cousin	14
						Cousin	14
						Cousin	14

Matched: 10/07/2017, Relationship: 3rd Cousin - 5th Cousin

(Ostrow Mazowiecka. Poland)| (Detroit, MI, USA)| (Ostrow Mazowiecka, Poland)| (Rochester, NY, USA)| (Ostrow Mazowiecka, Poland)| (Ostrow Mazowiecka, Poland)| (Canton, OH, USA)| (Detroit, MI, USA)| (Ostroleka, Poland)| (Ostrow Mazowiecka, Poland)| (Ostrow Mazowiecka, Poland)| (Biala Podlaska, Poland)| (Ostrow Mazowiecka, Poland)| (Ostrow Mazowiecka, Poland)| (Ostrow Mazowiecka, Poland)| (Baltimore, MD, USA)| (Ostrow Mazowiecka, Poland)| (Detroit, MI)| (Ostroleka, Poland)| (Ostrow Mazowiecka, Poland)

Image D. Family Tree DNA results, when viewed in the ADSA, provide more context, particularly in regards to ancestral surnames.

If you've uploaded data from Family Tree DNA, you can also (as previously stated) hover over the person's name and see additional information (image **D**). Specifically, you can view name, match date, estimated relationship, and estimated locations of ancestral surnames. These can provide valuable research tools that can connect you to genetic relatives. In fact, the results of this particular output were so promising that I stopped writing the book to e-mail this match—we have matching names and matching cities, which is quite rare in the Ashkenazi Jewish space!

I can additionally hover over the horizontal bar under the Segments column to learn more about the match date, estimated relationship, total shared segments, longest block, and segments by chromosome(s) and each of their sizes.

If you intend to use ADSA for triangulation, you should:

- Choose a long segment to research (at least 15 cM). If you're working with Family Tree DNA data, you can also choose a segment size as well as a surname by also selecting Highlight Matches with These Ancestral Surnames.

- Look for all segments that overlap—namely, those that occupy at least some of the same locations on the same chromosome pair.

- Compare all the ICW matches for individuals associated with these segments, and verify that all of these people match each other. This, in turn, will confirm whether they all are part of a triangulation group. (ICW matches, again, are predicted relatives, but may not necessarily share exact segments or strands.)

- Check the Ancestral Surnames to see if you and a match share a surname in addition to DNA.

- Reach out to your matches and work together to find that ancestor. You can also provide locations, as they could be more useful than surnames.

- Rinse and repeat.

All in all, the ADSA is a pretty interesting way to graphically represent your family, and you can send an e-mail and follow up with all family members on both Family Tree DNA and GEDmatch as the e-mail addresses are provided.

JWORKS AND KWORKS

DNAGedcom has two twin programs that operate similarly, but generate slightly different outputs. The pair, JWorks and KWorks, are intended to organize your match data so you can discover shared ancestors. The app sorts, groups, and builds matrices with your most promising matches. Both are available for PCs and Macs, and the only difference between the two is their output. (JWorks produces CSV files designed for Microsoft Excel, while KWorks produces CSV files designed to be opened in other programs.)

To use either of these programs, you must have already uploaded your Family Tree DNA, 23andMe, or GEDmatch Tier 1 data through DNAGedcom. JWorks and KWorks use that output for processing.

When you're ready to run JWorks, click "Click me to run JWorks." Add the kit number and change any defaults (e.g., you may wish to increase 6 cM to 10 cM) and let the app run. The process may take a while, and the output will be displayed as yourkitnumber_results.xls. Make sure to disable any security warnings on the top of your Excel application to both enable content and allow macros to run.

Your results (image **E**) are not particularly pretty or anything; after all, this is a spreadsheet. The output sorts, groups, and creates a matrix for your most promising matches. The names on the top rows are identical to the names on the left-hand column, followed

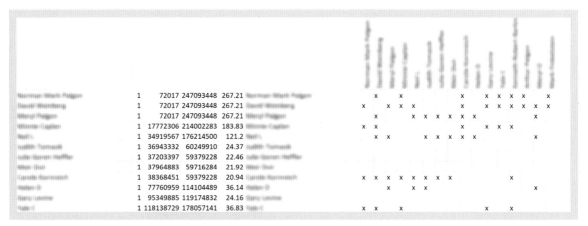

Image E. DNAGedcom's JWorks program provides you with an Excel-ready CSV file so you can better view and edit your DNA data.

by the chromosome number, start and end points, and size of segment. The names are repeated again, possibly just to make the matrix easier to access.

Similarly, KWorks, created by Kitty Munson Cooper **<www.kittycooper.com>**, takes your ICW data from Family Tree DNA (for example, if your kit number is B1111, you're looking for the B1111_ICW) and a segment CSV file (such as B1111-ChromosomeBrowser.csv). Click on Create Grouped File. It will take time to process, especially with lots of matches.

Genome Mate Pro

Genome Mate Pro (GMP) is a sophisticated application that analyzes data across a variety of platforms (23andMe, AncestryDNA, Family Tree DNA, and GEDmatch) to help you isolate common ancestors. Available for Windows, Mac, and Linux, GMP uses a single database to manage several kits at a time, allowing you to import data from the four aforementioned DNA testing services. The app also shows X-chromosomal (X-DNA) matches, features chromosome mapping for common ancestors, supports triangulation and ICW grouping, and allows you to import ancestors from a GEDCOM or Ahnentafel (ancestor chart) from AncestryDNA and Family Tree DNA profiles. In other words, it does it all. We won't go into great detail about using GMP, but we'll cover the basics. You can download GMP by visiting **<www.getgmp.com/download>**. From that page, you can also access the 290-plus-page GMP user guide, written by Jim Sipe.

The best way to use GMP is to upload GEDmatch data exports in addition to DNAGedcom files. Once you've created a profile, you'll need to upload your data. Click on Options, then Import Templates. To load your data into the system, choose the DNAGedcom files and go through the imports one by one—the exact files they ask you

GENOME MATE PRO
A Tool for Managing DNA Comparisons

Relative	Side	Paternal	Maternal	Relationship	cMs	Segs	Status	Date
	P			Great-Aunt	767.9	20	New	2017/10/23
	?				0.0	0	New	2017/10/21
	?			Distant Cousin	12.8	1	New	2017/10/23
	?				0.0	0	New	2017/10/21
	?				0.0	0	New	2017/10/21
	?			1st Cousin	754.8	0	New	2017/10/20
	?			4th Cousin	35.1	0	New	2017/10/20
	?			4th Cousin	36.8	0	New	2017/10/20

Image F. GMP provides you with a Relative List that describes how various matches are related to your DNA.

to upload. Note some files upload in minutes while some may take days. My Family Tree DNA ICW file took about 4.5 days to complete processing. Processing took far less time than my Ancestry ICW file, which was one-fifth the size because I was able to skip distant matches. Once you import your DNAGedcom files, you should then import your GEDmatch data by copying and pasting the information that will also be loaded to the app.

GMP lets you do quite a lot, even just from the main toolbar. Select a name from the dropdown menu at top left to switch between DNA kits/profiles. You can manage your profile or create new ones under Profile, and the Chromosomes tab allows you to view the chromosome browsers for all the relevant tests. The Relatives List allows you to view known relatives for all your profiles (image **F**), while Relative Detail allows you to drill down to specific relatives. Under Ancestors, you can load relatives and surnames from a GEDCOM file, and Segment Map allows you to view a breakdown of where you and a relative share DNA.

Eventually, you may be able to triangulate and learn more about other unknown matches. GMP takes a tremendous amount of data to make that happen. Going into details about the various triangulation functions are beyond the scope of this book, but the guide covers them in detail, especially as new features come out.

Double Match Triangulator

The Double Match Triangulator (DMT) **<www.doublematchtriangulator.com>**, a tool written by genealogist Louis Kessler and winner of the RootsTech Innovation 2017 award, works with autosomal data provided by Family Tree DNA, 23andMe, MyHeritage, and GEDmatch. The tool finds every segment from every matching person that double matches (i.e., who matches on both copies of a chromosome) and finding all of the triangulations between them.

To unlock the features of this tool as a Family Tree DNA user, you will need to pay for the Family Tree DNA Chromosome Browser on the accounts you need to test on, or you can hope the other person you're comparing against already has paid for the Chromosome Browser. You'll click on the Download All Matches to Excel. 23andMe users can do the same by going to the user's DNA Relatives page, then clicking Download Aggregated Data. On MyHeritage, you'll need to go to the matching relatives homepage, click Advanced, then "Export shared DNA segment info for all DNA matches." Finally, GEDmatch users will need to run a Tier 1 Matching Segment search. When this is completely done (remember, it will take awhile), you save the entire contents of this page to your clipboard. The, click Save GEDmatch under the File A section, and DMT will suggest a file name. With the comparison files in hand, this tool then analyzes the two chromosome browsers of match Person A and Person B.

To use the DMT, simply go to your chromosome browser, then indicate which file(s) or folder(s) you'd like to examine. The DMT produces an Excel file that maps every matching segment and lists all the people who share that match with the selected people whose chromosome browser files you have access to.

The results (image **G**) show how my results compare to the rest. Most entries are categorized as Full Triangulation, meaning there's a "double match," as Louis Kessler calls it. DMT finds every double match on the segments that involve the person being compared to. To put it another way: I match Person A, Person A matches Person B, and I match Person B all on the same segment (with significant overlap on all three chromosomes).

You'll need your own chromosome browser output, as well as another person's. The Full Triangulation designates a triangulation group, though some shared segment sizes may still be too small to be conclusive. Entries can also be indicated as "Missing a-b" (in which one entry doesn't overlap with one other) or "Base a-v."

The output may be difficult to read, so here's what each column means:

- **Name A**: The person whose test is input as File A (Person A).
- **Name B**: The person whose test is input as File B (Person B).
- **Name C**: A person featured in the match files of either Person A, Person B, or both.
- **Chr**: The chromosome upon which a match has been discovered.
- **Start-AC** and **End-AC**: The start and end base addresses that Person A matches with Person C (AC). Green indicates full triangulation: Person B also matches C and Person A matches B. Grey indicates a match between B and C, but not between A and B (indicated on the row as a Missing AB match). Red indicates the start and end base addresses start or end as a single triangulated match of A with C.

1	42260110	47629198	3.69	42260110	44380618	1.71	72017	247093448	01.000.247	Full Triangulation
1	42260110	47629198	3.69	42260110	44380618	1.71	72017	247093448	01.000.247	Full Triangulation
1	42260110	47629198	3.69	42260110	44380618	1.71	72017	247093448	01.000.247	Full Triangulation
1	42260110	46956783	3.05	42260110	44380618	1.71	72017	247093448	01.000.247	Full Triangulation
1	42260110	45252359	2.51	42260110	44380618	1.71	72017	247093448	01.000.247	Full Triangulation
1	42260110	44380618	1.71	42260110	44380618	1.71	72017	247093448	01.000.247	Full Triangulation
1	41918882	46956783	3.24	43219678	45252359	1.97	72017	247093448	01.000.247	Full Triangulation
1	43576450	47629198	2.38	43219678	45252359	1.97	72017	247093448	01.000.247	Full Triangulation
1	43576450	46600072	1.42	43219678	45252359	1.97	72017	247093448	01.000.247	Full Triangulation
1	43576450	45925790	1.32	42260110	44801407	2.24	72017	247093448	01.000.247	Full Triangulation
1	43576450	52923194	5.04	42260110	44380618	1.71	72017	247093448	01.000.247	Full Triangulation
1	72017	247093448	267.21	154848158	156486589	2.91	72017	247093448	01.000.247	Missing a-b Match
1	163398591	165765782	3.21	163646196	165765782	2.76	72017	247093448	01.000.247	Full Triangulation
1	164385968	169516601	6.84	164385968	169516601	6.84	72017	247093448	01.000.247	Full Triangulation
1	162870746	166684997	6.66	164755767	166684997	3.4	72017	247093448	01.000.247	Full Triangulation
1	165097600	171377576	7.68	165097600	171377576	7.68	72017	247093448	01.000.247	Base a-b

Image G. The DMT explains the relationships between me, a match, and another test-taker. For privacy reasons, this screenshot only displays columns Chr through Status; it omits the columns indicating kit numbers/names.

- **CM-AC**: The length in centimorgans between person A and C.

- **Start-BC** and **End-BC**: These columns compare person B to person C (BC). Green indicates full triangulation (Person A matches Person B and Person A matches Person C). Gray indicates a match with A and C, but not with A and B (Missing-AB), and blue indicates a single triangulated match with B and C.

- **CM-BC**: The length in centimorgans between person B and C.

- **TG-START** and **TG-END**: These give the start and end base addresses of the discovered Triangulation Group that was determined between Person A and Person C.

- **TG-GROUP**: The name of the triangulation group that was given when the DMT tool ran. This is based on AC matches that are included in the double matches with person B.

- **Status**: The status of the match, whether it is a Full Triangulated group; a missing AB match; Base AB (when Person A single matches Person B); or "C Is a B," which indicates that Person C is also Person B. Comparing all of the Person B's results can potentially help figure out if the matches are maternal or paternal.

To the right of the chart is a Map with color-coded regions to visually represent the matches. Green Xs Indicate double matches that triangulate (Person A matches Person C, Person B matches Person C, and Person A matches Person B). Grey Ms show double matches that are missing a match between Person A and Person B (Missing-AB). Red As show single triangulated matches where Person A matches Person C, and similarly, blue Bs show single triangulated matches where person B matches person C.

The Triangulator

The Triangulator **<dnagen.net/triangulator>** is another tool for Family Tree DNA users that triangulates autosomal DNA segments without going through third-party websites. It works within your browser, and we'll be examining how it works as a Google Chrome plug-in. When installed, the button appears as "dnagen tools" next to the ICW and Not-ICW buttons on your Family Tree DNA page.

With the Triangulator, you can choose up to five people to compare with your kit. In my case, I chose my mother, her brother, my first cousin twice removed (through my maternal grandfather's mother), my first cousin twice removed (through my maternal grandmother's father), and a third first cousin twice removed (through my maternal grandfather's father).

Image **H** is the initial result, which came out quickly. It estimates each person's relationship to each other, and (based on what I already knew) it was pretty accurate: My mother

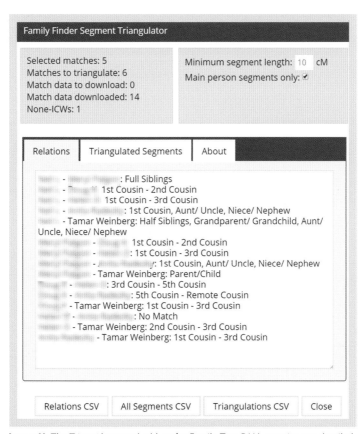

Image H. The Triangulator tool add-on for Family Tree DNA can give you detailed triangulation estimates. The Relations tab estimates the relationships between you and the matches you've selected.

and her brother are full siblings, my uncle and his father's maternal first cousin come up as first or second cousins, and my uncle and his maternal first cousin once removed are also in the first-cousin range. And then there's me, and he comes up as a possible uncle.

This cycle repeats for my mother and the three other individuals I compared against. On the third to last line, you see that there is no match—my first cousin twice removed on my grandfather's side (my grandfather's uncle's daughter—my grandfather's paternal first cousin) is not related to my first cousin twice removed on my grandmother's side. Two lines above that, another of my paternal relatives (first cousin twice removed, a.k.a. my grandfather's aunt's son) ends up having a fifth to remote cousin connection to the same woman. They're not related genealogically, but (because of endogamy) they share some DNA.

On the second tab (image **I**), you will see visualizations of the Triangulated Segments. This report breaks down individual segments that the individuals share, organized by chromosome and including which matches share that segment. In the image, you can see

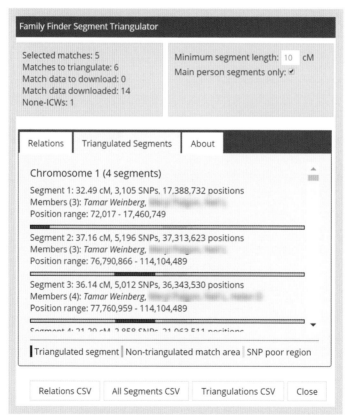

Image I. The Triangulator also provides detailed information about which segments you and your matches share.

that I have some reasonable overlapping matches with my mother and her brother, but on segment 3 my first cousin twice removed comes into play with a shared segment of 36.14 cM. The whole process repeats for all the compared relatives.

I can then download a bunch of CSV files, including Relations (the first screen), All Segments (which doesn't triangulate but shows shared segments between two people), and Triangulations (similar to the Triangulated Segments tab in the second image we saw).

RootsFinder

RootsFinder <www.rootsfinder.com>, a low-cost family tree software product, has DNA functionality that relies upon GEDmatch data. Using RootsFinder's tools, you can visualize your match data in interesting ways. We'll discuss two particular tools in this section.

GEDCOM AND BLOCK ANALYSIS

To use this first tool properly, you first have to upload a GEDCOM of your family tree. Then you will go to the DNA tab to import data from GEDmatch One-to-Many results. Simply copy the results onto the box and the results will be processed, creating a list of your matches organized by kit number. This chart shows the total number of cM shared, number of cM in the longest shared segment, genetic distance, total amount of X-DNA, and longest segment of X-DNA (image **J**).

Critically, the chart also shows the name of each match, along with a tree icon if that user has uploaded a GEDCOM to GEDmatch. You can add to this list without removing anyone just by tapping the + sign, then pasting results when showing the list of your matches. Over time, this makes RootsFinder's DNA tools more robust and comprehensive (though, as with many tools, less effective if you come from an endogamous culture).

This display is useful in and of itself, but you can do even more by adding family tree information. Select the kit, then click the pencil icon to edit information. If the person has not uploaded a tree but you recognize their name, simply hit Edit, then the Match tab. Here, you'll be able to provide information about who the match is. If you see someone has uploaded a tree, go back to your One-to-Many results on GEDmatch and locate that user's GEDCOM. Click GED on the results page, then the Pedigree link. From there, you'll be able to view relevant information about that person's ancestry.

Copy that, then go back to the edit match page on RootsFinder and click the Tree tab. Paste the results on the page, and you'll be able to see a list of that user's ancestors. If you recognize a match, go to the Match tab and choose the match from your tree. Remember, this is not very easy to do when there are unrecognizable ancestors if your ancestry is tied to endogamy, but other functionality is.

Once you've selected a match, click the arrow next to your kit number, then the little triangle from the dropdown to access Segments. These tools require Matching Segment

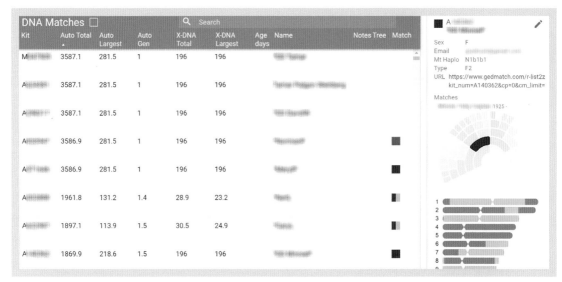

Kit	Auto Total	Auto Largest	Auto Gen	X-DNA Total	X-DNA Largest	Age days	Name	Notes Tree	Match
M	3587.1	281.5	1	196	196				
A	3587.1	281.5	1	196	196				
A	3587.1	281.5	1	196	196				
A	3586.9	281.5	1	196	196				■
A	3586.9	281.5	1	196	196				■
A	1961.8	131.2	1.4	28.9	23.2				▮
A	1897.1	113.9	1.5	30.5	24.9				▮
A	1869.9	218.6	1.5	196	196				■

Image J. You can paste a list of your GEDmatch matches into RootsFinder. You can also associate matching segments with individuals in your tree, allowing you to visualize a variety of information about them.

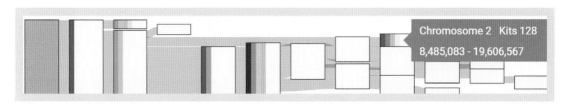

Image K. RootsFinder's colored block display allows you to see the relationships between the matches in your family tree.

Search data from GEDmatch, so you'll need a Tier 1 subscription to use them. Once you've run the test on GEDmatch and have your results, copy-and-paste the data in the RootsFinder screen. You may get a message that says Processing or see an error indicating whatever browser you're using has stopped being responsive. But don't cancel the processing—it just takes awhile.

Once your data finally processes, you'll see a page of blocks that represent a chromosome and segment matches with the kit (image **K**). Colored pairs represent matches shared by another relative that have already been introduced to the family tree. If you haven't linked up any listed family members, you will see plain white blocks.

TRIANGULATION

RootsFinder also offers a triangulation tool, and this is dependent on the Triangulation data under GEDmatch's Tier 1 plan. To begin using it, run the Triangulation data in GEDmatch. You'll want to change the view to list only the kit number, the chromosome, and the segment start position. Let the tool run for as long as it needs, and copy the entire

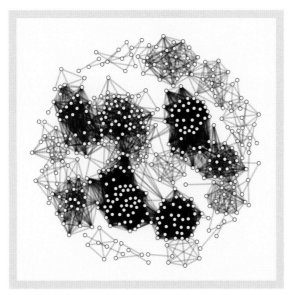

Image L. You can view the relationship between all your matches in a cloud using RootsFinder's triangulation tool. Each bubble represents a different DNA kit.

page's output to your clipboard. Return to the Triangulated section in RootsFinder and paste it to your page.

The tool generates a graphic that visualizes how you and all your matches are related, organized into distinct clusters (image **L**). However, if you come from an endogamous culture, you will have a mishmash of a single cluster (and maybe some others on the side) that looks like the image here showing my triangulated results with the defaults (way to go, Ashkenazi Jewish endogamy!).

The full image shows how your matching kits relate to one another, with clues about how you connect to each other and color-coded branches for identified family members. Each dot represents a different match, and you can hover them to learn about that person and your chromosome matches. Shift-click allows you to drag clusters to show details for each kit in the cluster. You can also play with the relationship range to see your triangulated groups become bigger or smaller.

If you see distinct clusters, these represent groups of closely related people. Bigger clusters (more bubbles) represent a lot of test-takers who are closely related. Smaller clusters with fewer bubbles may require more test-takers to ascertain the common ancestor. But if you identify cousins in each group, you can collaborate and may be one step closer to figuring out how you are all related.

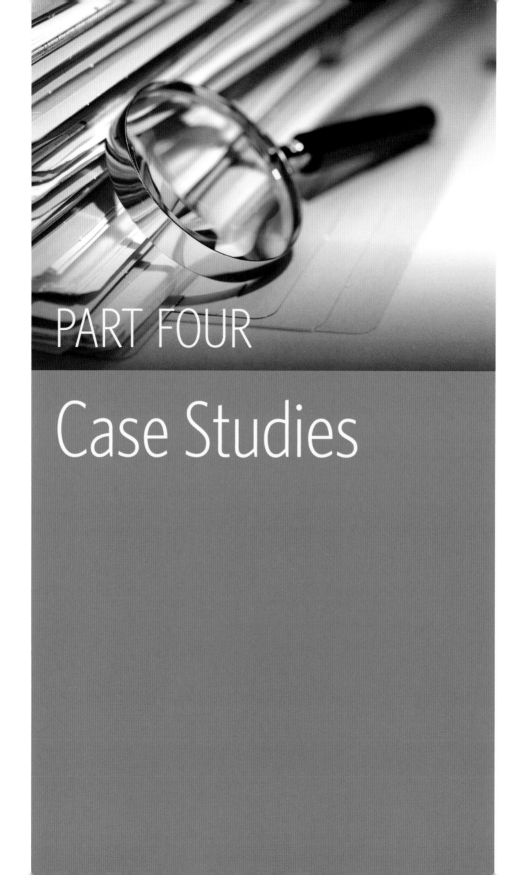

PART FOUR

Case Studies

CASE STUDY

Donna

L ike me, Mark is a genetics geek who is always coming up with new discoveries for his family tree and the trees of his extended family. My great-aunt married Mark's uncle, so we're related by marriage. Surprisingly we also share a lot of DNA, most of which we may never trace back to a recent ancestor because of our endogamous community. Mark and his wife, Jaye, shared with me the story of how they helped a genetic match named Donna, who was somehow related to Jaye's family.

Jaye's brother, Jim, had a high-DNA match named Donna, with 160 cM total shared DNA and one segment over 40 cM). Donna was searching for her biological father; her mother's husband could not have children, and so her mother conceived her via artificial insemination. However, Donna couldn't find any records of her birth. She knew the location in Westchester County, New York, where she was born, and she had reason to believe the donor was a doctor or medical resident. She assumed that he would have to be at least in his late seventies by now.

By viewing Jim and Donna's shared cousins on GEDmatch <www.gedmatch.com>, Mark was able to determine that Donna and Jim matched through Jim's maternal grandmother's side, which included the surnames Ruttenberg and Gelfond. Most of this family immigrated to Boston from Ukraine in the late 1800s. Donna confirmed that she must be related to Jim through her birth father's family; her half-brother, who had the same mother but a different sperm-donor birth father, tested and didn't have Jim or Jay as matches.

Given that, Mark was able to limit the potential fathers to Jaye and Jim's maternal cousins, many of whom they knew of thanks to previous research. Since the insemination took place in New York City in the 1960s, Mark singled out a single male cousin who would have been the proper age and in the proper place: a biochemistry doctoral student named Michael who moved from Boston to New York in the early 1960s. Michael is Jim and Jaye's first cousin once removed, and he died in 2000.

Intrigued, Donna reached out to Michael's former girlfriend, who confirmed Michael had talked about being a sperm donor in the 1960s in New York. The hospital had marketed extensively to doctoral students and residents, and donors could earn up to $75 per week—good money for students. Apparently, Michael wondered if he would ever meet any of his children.

So Mark had strong circumstantial evidence that Michael is Donna's biological father, but what other clues could he discover? With 160 cM, Jaye and Donna were estimated to be in the second-cousin territory. With no other relatives to verify (Michael didn't have any other children), Mark and Donna turned to GEDmatch to put the story together. Doing this, they found that a few of Donna's closest matches all fit within Jaye's family (image A).

Let's look at the matches who shared the most DNA with Donna. Donna's closest match was Jim. The second highest match was a man named Robert, who descends from Michael's great-grandparents (Jim's great-great-grandparents), making Robert and Jim second cousins once removed. The third match was named Yefim, who (as it turns out) was Jim's closest DNA match. Mark reached out, and found out a branch of Yefim's family was from Berdihev, Ukraine, where Jim's maternal great-grandfather and Michael's paternal grandfather originated.

Yefim was a shared match between Donna and Jim, reaffirming the relationship between Donna and Michael (who we know from genealogical research is related to Jim). This suggested the two shared a common ancestor in one of Michael's great-grandfather's parents. And since Yefim matched Jim and Donna but not Robert, we can further narrow which branches of the family tree might harbor the shared relative.

To further confirm the hypothesis, Mark looked into Michael's mother's side, as the matches to date all were paternal. The GEDCOM database at GEDmatch had no surname matches, so Mark visited Ancestry.com <www.ancestry.com> to find the surname in its

					Haplogroup			Autosomal				X-DNA	
								Total cM	largest cM	Gen		Total cM	largest cM
Type	List	Select	Sex	GED/WikiTree	Mt	Y	Details				Details		
					▼ ▲	▼ ▲		▼	▼	▼ ▲		▼	▼
F2	L	☐	M	GED		J-M267	A	158.9	32.3	3.2	X	0	0
F2	L	☐	M				A	155.6	46.9	3.3	X	0	0
V4	L	☐	M	GED	H1e	R1b1b2a1a	A	102.5	22.8	3.6	X	0	0

Image A. Using GEDmatch, Donna identified three DNA matches—Jim, Robert, and Yefim—who also fit into Jaye's family tree.

database. Putting it all together, Mark is fairly certain by following all these ancestral lines on the great-grandparent side that he's discovered a new second cousin. It's still a work in progress, but Mark is confident he's on the right path. DNA, after all, is always a work in progress.

Mark and Donna continue to look for answers. Donna uploaded her results to multiple testing sites, and MyHeritage DNA **<dna.myheritage.com>** predicted one match was her half-sister. Upon reaching out to her, Donna learned she was born in the same hospital where Michael had been enrolled as a sperm donor, two years after Donna had been born. This woman also matched Jim and Jaye as a second or third cousin, consistent with what Donna's half-sibling would be.

Despite the evidence, the match denied the relationship. As far as she knew, her mother hadn't been artificially inseminated. She insisted the man who raised her was her biological father, and that the test must be mistaken. Donna didn't force the issue—not everyone wants to know the truth about their past.

Donna acknowledges that her struggle is more difficult than others searching for donor biological fathers from her time period, because no records were kept. She finds it difficult because each state has variable laws about whether donors can be anonymous or if there's a limit on donation. Of course, she is hoping to find those answers soon, and her search is ongoing. She and Mark continue asking relatives to test, adding more and more evidence.

CASE STUDY

Izak

zak and his brother, Shep, were separated after World War II. With the assistance of MyHeritage, the brothers found closure—after waiting nearly seven decades to do so.

Born in Bergen-Belsen, Germany, in a displaced persons camp after World War II, Izak found himself in turbulent times, a period that was defined by murder and genocide. Fortunately, Izak did not remember much about his formative years—but does remember life after moving to Israel where he was adopted at age three in 1948. He knew his biological mother, Aida, relocated to Canada, and they connected and maintained contact in his teenage years. However, Aida kept most of her family secrets to herself; Izak knew nothing about his father or anyone else he may be related to.

With the help of documentation provided by the Bergen-Belsen archives, Izak made a shocking discovery: He had a brother (named Shepsyl) and another relative—unknown at the time—at the camp. Both had relocated to Canada as well, but not with his mother. Naturally, questions emerged: Was his brother alive? Who was the unknown relative, and could it have been his father? He had to find out.

With information in hand, Izak's nephew, Alon Schwarz, reached out to MyHeritage to see if they could help Izak find his brother. With the help of MyHeritage's founder and CEO Gilad Japhet and its senior researcher and Head of Genealogy, Laurence Harris, Alon and the MyHeritage team perused documents all over the world: at Bergen-Belsen, Yad Vashem <www.yadvashem.org>, online family trees, and vital records from Canada and Israel.

Laurence eventually stumbled upon a woman named Melanie, who was the daughter of a blind Canadian man named Shep. Melanie told him that Shep's name was Anglicized from Szewlewicz to Shell upon immigration to Canada, and he was born in a displaced person's camp in Germany. All signs pointed to this being Izak's brother, and it all made sense.

The brothers reunited—or, rather, met for the first time—in Canada. Izak learned fascinating things about his younger brother. For example, Shep had been a visually impaired Paralympic skier, cyclist, and marathon runner. The two later visited Aida, putting the family back together.

Alon, a producer, wrote a documentary about the story, called *Aida's Secrets*. The film goes into detail about Izak and Shep's story, and it was launched in limited release in October 2017. You can read more about Izak's story on MyHeritage's website **<stories. myheritage.com/izak-szewelewicz>**.

CASE STUDY

Jack

J ack (name changed) discovered his mother was adopted. His story hit home for me because he's my cousin. I discovered him when I was going through my grandmother's matches and saw a third cousin match I never heard of—and whose family names did not look remotely familiar (nor Jewish). Baffled, I reached out to him and started a conversation. It took awhile for him to open up, and I know now it was because of the sensitive nature of his familial situation.

Jack's mother is of Asian descent, while his father was said to have Colonial American roots. However, Jack's DNA showed otherwise. First, Jack was predominantly European Jewish—close to 50 percent. (The rest was an expected mix of East/South Asian and Polynesian DNA which was clearly maternal.) Jack's story, then, seemed to stem from a "non-paternity event," in which some action (infidelity, adoption, etc.) results in an unexpected break in a genetic line.

I believe this news came as a shock to Jack. He suspected something was awry with his heritage, and his mother denied any peculiarities. Once he told me his predicament, I vowed to come up with an answer, if only one day. I imagined it was emotionally hard for him to try to solve the mystery, so I took it upon myself to find the answers for him.

I discreetly used the information I had to find more. I already knew when and where Jack was born, and I had a potential picture of where a father could have been at the time of Jack's conception. In fact, I knew relatives of my grandmother's who lived in the same area at the time, narrowing possibilities for me. A group of three brothers, last name Lawrence (name changed), was interested in genealogy, and I wondered if they might provide answers. Perhaps Gary, one of the Lawrence boys whose personal timeline fit in with Jack's conception, could be the missing father.

As luck would have it, a Gary Lawrence happened to test his DNA and showed up as a match—a second cousin of my grandmother. I immediately told Jack about this potential match, but Jack and Gary were estimated to be just fourth to sixth cousins. My hopes were shattered—for about a day. Then I contacted Gary and learned he was *also* adopted, and I realized I had the wrong guy. Same name, wrong Gary Lawrence. What are the odds of that?

I now had two mysteries to solve: Who is this "wrong" Gary? He came up as a first-cousin match with another cousin of mine, Alan. Alan had one aunt and two uncles who were candidates as Gary's parents. Based on Gary and Alan's ages, I ruled out the aunt—Alan would have been old enough to have known if his aunt was pregnant and put up a baby for adoption, and (in any case) Gary already knew the name of his birth mother. That left the other two brothers, one who had kids and one who didn't.

First, I researched the cousin who didn't have kids. I left him a message under the guise I was doing family tree research (which was, in fact, quite true), and he called me back, immediately telling me that he doesn't know much about the family and had moved to the other side of the country anyway. Still, I shared what I knew to get comfortable with him. (After all, he was family!) I asked some questions related to Gary, and it didn't take long for me to figure out that he was, in fact, Gary's father. All I needed was to name Gary's birth mother, and he confirmed it all. He said he and Gary's mother were young and in school, and both just weren't prepared to be parents. Neither of them forgot about their son, and when they'd talked about him from time to time when they reconnected over the phone.

With that settled, the "real" Gary Lawrence still remained a mystery. I knew his two brothers, who had both been participants in family genealogy (though not in DNA research). After all, no match came up for them in any of the databases I had my data connected to—except, finally, for one day when a Lawrence brother, who we'll call Fred, showed up on 23andMe **<www.23andme.com>**. I immediately reached out to him and told him to import his data to GEDmatch **<www.gedmatch.com>**.

He complied, and I ran a one-to-one between Fred and Jack. *Voila*—3,585 cM shared. Fred is the father. So my initial guess wasn't exactly accurate—I had the right family, but the wrong brother. I got on the phone with Fred, who confirmed he had a relationship with a woman around the time of Jack's birth and learned she was pregnant thereafter. Not knowing who the father was, he didn't pursue it, and he didn't think Jack was his child anyway since he learned about the pregnancy in passing. When he found out Jack was his son, he was surprised—sad for missing lost time (and angry for not having been told the truth sooner), but excited to meet this new part of his life.

Jack grew up with a father he thought was his own. It was only later that he suspected otherwise, and a simple DNA test connected him to his father—and the "wrong" Gary

Lawrence to his father, as well as me to all of them. With that one text, we were able to solve a number of family mysteries. It took time, effort, and testing on different platforms. It took cooperation from everyone, and it allowed all parties to gain some closure.

CASE STUDY

Kalani

K alani is Kānaka Maoli (native Hawaiian), an ethnicity that many testing services show as Polynesian. His mother, Judy, was adopted. Both of her parents were, as she understood it, Hawaiian. In an effort to find out more about herself, she took a DNA test in May 2013. However, as you may recall from chapter 3, Polynesians are an endogamous culture—no matter what new communities were founded in the area, geographic isolation still had communities connected to the same few ancestors. This is especially true of New Zealand (Maoris) and Hawaii (Hawaiians), where Polynesians were last known to have settled.

As a result, Kalani ran into similar challenges that Ashkenazi Jews face. These communities didn't use surnames until recently (approximately 1860), and the surnames that *were* adopted were often inconsistent between a family on one island and their relatives on another. For Kalani, it was not going to be so easy to find a biological connection for his mother.

He turned first to Family Tree DNA <www.familytreedna.com>, however endogamy prevented him from finding relevant matches. He had five pages of matches of individuals who shared more than 300 cM, but none of them had longest segments of more than 20 cM, rendering most of the matches useless. And because Family Tree DNA uses small segments in its default results, tiny segments often contribute to inaccuracies in relationship predictions. In other words, an assumed first or second cousin is likely not one at all. Kalani had to be mindful of that as he reviewed his matches.

With no promising match, Kalani decided to test on 23andMe <www.23andme.com>. But even there, he had dozens of predicted second cousins. Even when importing the data to GEDmatch <www.gedmatch.com>, he had pages upon pages of "significant" results, with the highest at 510 cM: eight matches in the 400-cM range, and over a hundred more in the 150- to 300-cM range.

Kalani ventured off and tested on AncestryDNA **<dna.ancestry.com>**. His closest matches fell within a range of 187 cM and 304 cM. He felt discouraged without a chromosome browser on the service, especially since the patterns seemed to reproduce each other on every single DNA testing platform with many high matches and not enough to work off of, and he wasn't able to establish the true connection between any of AncestryDNA's matches. As such, he ignored the results for a year.

Still, Kalani wasn't the type to give up. He knew he'd have to work harder to find someone closely related to his mother. After some time passed, Kalani found a Family Tree DNA match who shared 266.94 cM and a largest segment of 50 cM with Judy. With a segment of that size, the relationship was likely close enough to matter. The match cooperated and shared her family tree, and her ancestors included a Swiss man (Alfred Manner) and a 100-percent Hawaiian woman named Kama'u.

This presented Kalani with some possibilities to explore. Kama'u had a daughter named Theresa, and Theresa's husband was born near the same area as Judy. He put a pin in this lead for the moment.

Kalani returned to AncestryDNA, where Judy's top match belonged to a man whose tree contained unfamiliar names. Kalani chalked this up to endogamy yet again, with the assumption that the match wasn't as close as he thought.

Judy, while predominantly of Polynesian and East Asian (Oceanic) ancestry, also had a relatively small percentage (17 percent) of European ancestry. This led Kalani to search for a European ancestor who would have married a Hawaiian. Judy recounted meeting her biological father at age 5 and felt sure he was 100-percent Hawaiian, so Kalani ruled out the possibility of him being all or mostly European. Further, Kalani's mtDNA haplogroup is B4a1a1a3, a subclade of the Polynesian haplogroup B4a1a1. Given the haplogroup's geographic isolation, he determined his direct maternal line (and, thus, Judy's maternal line) was Hawaiian.

Kalani started working toward his potential European ancestors by creating a chart of what possible ethnicities his ancestors could have been (image **A**). Using 23andMe's X chromosome browser, Kalani observed that his mother had an X chromosome with no ethnically European DNA on it. Given the X-DNA inheritance pattern (see chapter 4), Kalani was able to identify where in his genetic family tree Judy's European X-DNA might have come from. He estimated an ancestor around the 1860s was 50-percent European and 50-percent Hawaiian. With Theresa, the daughter of Alfred Manner (a Swiss man) and Kama'u, being born in 1866, he knew that he was getting closer.

Still, a 50-cM segment may indicate a shared ancestor further back in time. Kalani had a Family Tree DNA match (Theresa's husband) whose father's Y-DNA results showed

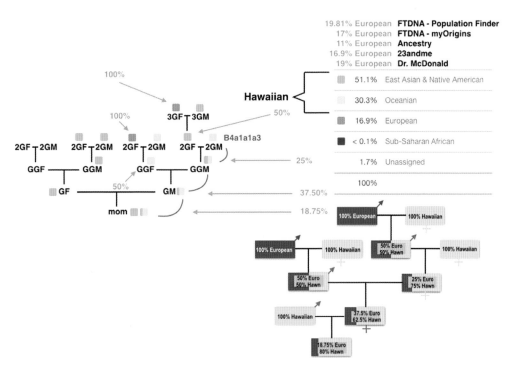

Image A. Kalani created a diagram to determine how his mother, Judy, inherited her ethnicity along X-DNA. To do so, he had to study X-DNA inheritance patterns and view Judy's X-DNA data in a chromosome browser.

European origin—not Polynesian, as he expected. Knowing this, he figured a non-paternity event (NPE) occurred somewhere because he expected a 100-percent Hawaiian connection. This eliminated the possibility of Theresa's husband, their children, and her husband's siblings and children as the source of the European DNA. That ruled out everyone in that family group besides Theresa herself.

But the Theresa lead wasn't totally dead. Perhaps Kalani and his mother were related to one of Theresa's or Kama'u's siblings? But then, Kalani made another discovery: Theresa had a first husband: John Holbron from England. The name Holbron stood out because it was tied to the tree of his top match on AncestryDNA. Now, things were starting to come together. Theresa and John had a son named Robert—and that's when Kalani hit the jackpot.

Referring back to his diagram, Kalani determined this Robert was the most likely source of his mother's European DNA. Robert married a woman named Annie, the daughter of Ehu (who was half-Hawaiian and half-European) and William Ludlum, a full-blood European. Ehu, he assumed, could have carried the B4a1a1a3 haplogroup throughout the generations. Later on, Kalani would find Robert and Annie's estimated years of birth fit his diagram, and the predicted year of birth was off by merely a decade. Now, Kalani had a potential ancestral line to work off of, beginning with Robert and Annie's children and grandchildren.

From the Holbron match on AncestryDNA, Kalani found Rose Holbron (Robert and Annie's daughter), Frank Kanae, and their three daughters. Their tree included photos, and two of the subjects in one of them looked just like Judy and his sister. While researching Rose's family further, he found that Rose had given birth to a child just one month before Judy was born, so Rose could only be Judy's birth mother if the birth certificate Judy had was incorrect.

Judy's birth certificate did not acknowledge that she was adopted. She was born in a residence and delivered by a midwife, and the certificate indicated both (adopted) parents permanently resided at a different address. Kalani looked into these locations thoroughly but did not make much headway.

He turned his efforts back to Rose and Frank's three daughters, one of whom, as we stated earlier, looked like his mother and sister. One of the other daughters, born Rose Kanae, married several times and had nine children of differing surnames (likely to be from her multiple husbands): a man Kalei, Joseph Akana (a name that could be either Hawaiian or Chinese), and one other man.

Kalani found that Joseph Akana lived at the address on the birth certificate. Does this mean that Joseph Akana is Judy's father? Not necessarily. Rose could have been impregnated by another man, but had the baby in her husband's home. Or perhaps one of Rose's daughters got pregnant while Rose was married to Joseph, and gave birth at her parent's residence.

Around this time, Judy's court petition came back with some non-identifying information about her ancestry. The documents stated Judy's parents were both Hawaiian and Chinese. Chinese? Her ethnicity findings didn't support that at all. Still, it showed a strong likelihood to Rose Kanae and Joseph Akana as being the parents (though Joseph may not be the biological father due to his confusing surname). Because of questionable paternity, Kalani believed that Judy's mother may have remarried.

So why were Judy's parents listed as being as Hawaiian and Chinese? Back in the day, regardless of state, if a mother is unwed and putting her child up for adoption and knew the name of the adoptee's mother, she may take on the identity of the adoptive mother. The woman who adopted Kalani's mother was, in fact, Hawaiian and Chinese, potentially explaining the inconsistency between Judy's adoption papers and her ethnicity estimates.

Despite this, Kalani explored his ancestral lines as if Rose Kanae were Judy's mother (and his maternal grandmother). Since Kalani already knew that, based on his mtDNA

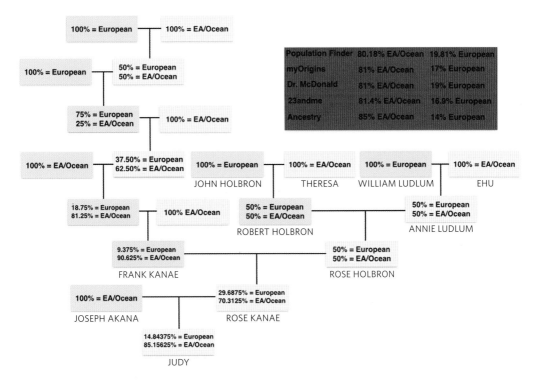

Image B. Now that Kalani believes he's found his mother's birth parents, he re-created an ethnicity family tree and compared it to what he knew so far. The percentages mostly match up, providing evidence that his theory is correct.

haplogroup, his maternal grandmother was of Hawaiian descent, he came up with four men who could have provided his and his mother's European DNA:

1. A man named Oliver, who was American and settled in Hawaii in 1793

2. Oliver's son-in-law, Isaiah Lewis

3. William Ludlum, the European man mentioned earlier who married Ehu

4. John Holbron, Theresa's first husband.

With a deeply rooted family tree, Kalani created another diagram using Keynote **<www.apple.com/keynote>** with potential ethnicity percentages for each ancestor (image **B**). If Joseph Akana and Rose Kanae are Judy's parents, the percentages more or less line up.

Kalani next searched newspaper clippings for birth announcements around the time his mother was born. In one, he found a birth announcement for a Joyce, the daughter of Mr. and Mrs. Joseph Akana. Joyce was Judy's original name! He finally had a paper record to support his new theory.

He shared his finding with Judy, only to discover she already knew much of this and even recognized the name Akana. As it turned out, Judy withheld some information from her son—an all-too-common occurrence in adoption cases, those involved might not want to fully disclose all the available information.

With that, Kalani reached out to his and Judy's new relatives. He gathered information from the obituaries of Judy's siblings and half-siblings, and he reached out to them with his findings. The photos they shared with them were uncanny—Judy shared a resemblance with many of her siblings. Kalani also used Facebook's mutual friends trick (explored in chapter 2) to get in touch with another cousin who (as it turned out) had known Judy for years.

After uniting with his family, Kalani learned more about Rose's history and her three marriages. Of the nine children, all but two were given up for adoption. The elder siblings sort of knew about Judy's existence—they had heard of a woman named Joyce who was adopted into a Filipino family. (Indeed, Judy's adoptive father was Filipino.) None of the other siblings had known where Judy lived or what had become of her.

Finally, Kalani gained closure on the suspicion that Joseph Akana was really his birth grandfather. As it turned out, Joseph adopted the surname "Akana" later in life. He was born Hawaiian (and was just as Hawaiian as Judy remembered him being when they briefly met when she was five), but members of his generation tried to downplay their Kānaka heritage. The Kānaka from his generation could be punished for speaking Hawaiian, and so Joseph wanted to appear as something more generally accepted. He chose "Akana" because it was an ethnically ambiguous name.

Kalani has since tested several of Judy's relatives, including her half-sister, a nephew, a half-niece, and a first cousin's daughter. All confirm his previous findings. He was even able to test some of Joseph Akana's relatives once he confirmed Joseph's original surname was "Kaapuiki."

Kalani tried all the tricks in the book (specifically, in this book), and it paid off thanks to his patience and perseverance.

CASE STUDY

Kelly

All her life, Kelly wished she had a father. She worried about father-daughter dances and yearned to have someone walk her down the aisle if she got married. Who would playfully intimidate suitors? Who would cheer for her at softball and volleyball games? She tried setting her single mother up on dates, but she wasn't interested in dating. Without a father, Kelly always felt so alone, especially when so few other children came from similar backgrounds.

Perhaps that's why, when she was little, Kelly believed that women could wish themselves in and out of pregnancy. They could just decide that they wanted to be pregnant, and would spontaneously have a baby nine months later. After all, when Kelly asked her mother (as all children do) how she had her, her mother simply said she prayed for her.

It wasn't until the third grade that she learned the truth. When maiden names came up one day in class, Kelly's Spanish teacher couldn't understand why Kelly's last name was also her mother's maiden name. The teacher seemed to find it offensive, and scolded Kelly in front of her peers. Upset, Kelly hid under a desk and sobbed. A classmate came to comfort her, and shared her theory: that Kelly had her mother's "maiden" name because she had been conceived via an anonymous sperm donor. Kelly didn't know what this meant at the time, but (lo and behold) her mother confirmed that was the truth.

Kelly became used to explaining her familial situation, and thought she was okay with being donor-conceived, even if she still felt a longing for a father. But as she grew up and changed schools, there were more and more people she had to explain her situation to. Most thought it was "cool," which was strangely unsettling for Kelly, but one acquaintance clicked her tongue and said, "That's wrong."

That sentiment stuck with Kelly over the years, and it made her want to better understand where she came from. In science classes, she learned about the mechanics and biology of reproduction, but she still knew so little about where her own genes came from. She spent hours on the school library's computers, researching issues related to sperm donation. In the days before Facebook, she found Yahoo forums for people related to sperm donation—including communities for donor-conceived person(s)/people (DCP), as well as recipients and donors.

She made several friends in the DCP community. They were all searching for the same thing, and all those special links made them feel like they were part of a family.

From these groups, Kelly found other resources for learning about her genetic father—but encountered several barriers. First, learning about her family would cost her a significant amount of money. Consumer DNA testing was still in its early, expensive stages. And since she (as a DCP rather than a donor or recipient) wasn't considered one of the company's patients, privacy restrictions stopped her from being able to access her own medical history.

She did make one big breakthrough, however. Kelly found the Donor Sibling Registry <www.donorsiblingregistry.com>, an organization dedicated to connecting DCPs with their birth siblings. Through the registry, Kelly discovered that each company assigns IDs to donors. More significantly, she discovered her birth father's ID was 007 (though she felt certain her birth father was not James Bond).

Kelly received eight full pages of information about 007, unusual for an "anonymous" donor. He evidently thought the information wasn't identifiable when he consented to providing it all those years ago, but now (armed with Google), Kelly could try. The papers stated her birth father was a professor who received several degrees at different universities, but she couldn't find a professor with that set of credentials. Could she trust the information here? She would have to conduct further research to find out.

Kelly became active in other ways as well, encouraging the *Houston Chronicle* to run a piece on the importance of telling children their true biological origins. Part of the problem, Kelly figured, was that many DCPs (including her potential half-siblings) didn't know the truth about their birth. Kelly's mother, wanting to keep the matter private, threatened to disown her if she attached her name to the article, and so Kelly had to abandon it since the paper wouldn't print a story anonymously. Eventually, she decided to take time off from her efforts to study psychology in college.

Once she graduated from college and had "big-girl money" a few years later, Kelly took her first DNA test through Houston-based Family Tree DNA <www.familytreedna.com>. Despite her excitement to make DNA discoveries, Kelly couldn't find any close matches through the site. She knew she needed to take tests with AncestryDNA <dna.ancestry. com> and 23andMe <www.23andme.com>, the other big testing companies at the time.

After a couple years of pinching pennies to afford these other tests, Kelly still had no close matches. Instead of half-siblings or parents, she found a lot of fourth, fifth, and sixth cousins. Her ethnicity estimate provided some interesting information, however—she was one-third Ashkenazi Jewish. Raised United Methodist in a majority-Jewish neighborhood, Kelly was surprised. She had attended bar and bat mitzvahs growing up, but that was the extent of her exposure to Jewish culture. She was sad she had been so close to her heritage without knowing it. She also felt cheated because the papers from her birth father's 007 file stated he was Protestant, not Jewish. Why would those records (which, after all, were from a licensed medical establishment) lie to her? Without reliable information or close DNA matches, Kelly didn't know how she could connect with her birth father or siblings.

Kelly then heard about the DNA Detectives Facebook group <www.facebook.com/TheDNADetectives>. Here, volunteers help adoptees, DCPs, and others who were stolen or otherwise separated from their birth parents reconnect with genetic relatives. Using shared matches, available records, and other key resources, these volunteers could work miracles for people like Kelly. And, best of all, they worked for free. After just a few weeks of waiting, the DNA Detectives paired Kelly with James (a DCP's husband) and Jennifer.

The pair worked for several months, and eventually a candidate emerged from two fourth-cousin matches: Keith. His documented family tree (which the pair and Kelly created using obituaries and other publicly available records) included common Jewish last names like Weinberg(er) and Goldberg, and Kelly was able to find that he owns a tourism company in addition to teaching history and Judaism at a college. She even found reviews from his students and customers, some of which weren't kind. She took this commentary with a grain of salt—after all, most people only write reviews when they're angry.

Armed with an e-mail and home address, Kelly was ready to take the next plunge and contact Keith. She read over threads on DCP Facebook groups that discussed best strategies for writing letters to donors. Like many DCPs, Kelly wanted to make sure Keith didn't misunderstand her motives—she didn't want money or to impose a relationship on him. She just wanted answers about her heritage.

On July 4, Kelly wrote to Keith for the first time—her own personal Independence Day. In it, Kelly introduced herself as a DCP and summarized her research thus far (including the donor ID she'd received from the Donor Sibling Registry). She explained that, based on shared DNA matches and the information from the eight-page packet she'd received, Keith was likely her father. After years of searching, she simply wanted confirmation and to reach out in a way that was comfortable for both of them. She provided contact information for herself and offered to have a third party speak with Keith before they made direct contact. She also gently asked if he would be interested in taking a DNA test.

Kelly didn't have to wait long for a response. The next day, she received an e-mail while at work. In it, Keith confirmed he had been a sperm donor and that he knew he had several children born to various mothers around the time of Kelly's birth. He even offered to potentially meet her someday (though Kelly wasn't quite ready for that yet). Overjoyed that she had potentially found a father (and that she had half-siblings), Kelly shared the good news with her husband, Jennifer, and James. She waited a few days to respond, continuing what would become a month's-long correspondence.

In his e-mails, Keith seemed to have just as many questions about Kelly as she did about him. He asked how she had found him and learned about his family tree—what happened to her mother, and what kind of life she had growing up. Kelly didn't volunteer information, but was happy to answer questions all the same.

As it turned out, Keith shared Kelly's interest in genealogy, but the conversation took a turn for the worse once Keith began discussing his family history. After innocently asking what had gotten Keith interested in genealogy, Kelly received detailed stories about Keith's dysfunctional childhood. He did not have the best of relationships with his brother or mother, who he said married five times. He only saw his father briefly several years after his parents' divorce. He shared more of these uncomfortable stories in a similar vein.

Kelly didn't know how to respond. She'd only asked questions that related to Keith's interest in genealogy, but he still disclosed a ton of information about his family's history that Kelly didn't ask for. Was Keith's oversharing jeopardizing their relationship? And how could she reconcile the discomfort she felt with the father she had spent years searching for? Throughout their correspondence, Kelly kept trying to divert the conversation to another topic entirely.

And that's when their correspondence hit rock bottom. While Kelly found the closure she needed, it didn't end on the happiest of notes. Although Keith had candidly shared his memories of his dysfunctional family (and despite Kelly's efforts to steer away from that topic), he suggested that Kelly's questions were drudging up too many unhappy memories, and told her to leave him alone altogether.

Kelly was devastated. In his own words, Keith felt like an orphan in his dysfunctional family, and had seemed to welcome her as a fresh start. But despite trying to giving him clear and healthy boundaries, Kelly felt she had been unjustly rejected. She didn't know where she belonged and fell into a deep, albeit secretive, depression. She tried to stay busy with work, and her husband and DCP friends tried to cheer her up. Still, she nearly gave up.

Kelly sought out therapy and threw herself into her work at Meals on Wheels, where she found purpose and a supportive community. As part of her job, she communicates with hundreds of people each month, each sharing their own stories and reminding her that she isn't alone.

Kelly also continues to find community in her new DNA matches. Each new match further suggests that Keith is her father, but her ability to connect with other cousins helps her find a sense of belonging. One cousin, Joel, even offered to "adopt" Kelly and have her come visit him in New York, a suggestion that deeply touched Kelly. She also got in contact with a woman who matched her as a sibling, though she wasn't interested in pursuing a relationship. Still, Kelly was reminded that she wasn't alone.

Kelly hasn't attempted to contact Keith again, save for a quick message to make sure he was safe after Hurricane Harvey. (He was.) Despite their last, painful e-mail, Kelly doesn't regret finding Keith. She received the closure she needed, and she's grateful to close the door on that major question mark in her life.

CASE STUDY

Marcy

When Marcy was two weeks old, a loving family adopted her. The adoption was never kept a secret; her adoptive parents always told Marcy that her birth mother was single and couldn't take care of her. Thus, her birth mother placed her with an agency to give her a better life. Every year on the anniversary of her adoption, Marcy "adopted" a doll to celebrate. Since she had always known she was adopted, she used to joke that her parents stood over her bassinet every day and told her she was adopted.

Still, Marcy had questions. No one in her adopted family ever treated her differently, but she knew that their family histories were not her own. Her parents gave her a good life, but Marcy's teenage years were very turbulent. She often told her adoptive parents that her "real" mother would understand her. Her adopted family was not perfect, but not awful either. Marcy thinks that's true for most families, whether they are birth families or adopted families. Marcy still wanted to know more about where she came from.

Her search started with just a name. Marcy's adoption papers list her birth mother's name—it had never been a secret to her. She used to look through phone books to try to find her birth mother's name. She asked everyone with her birth mother's last name if her mother was a family member.

Beyond that, Marcy couldn't do much back then. When she learned that the adoption agency would send her some information, she submitted a request and received non-identifying information. She also registered with a few places to see if anyone was looking for her. Nothing ever came of it, but Marcy still believed that, unless her birth mother was dead, she would be looking for her. Even Internet searches proved fruitless.

During all this time, Marcy was raising her own family and enjoying the special moments. Her adopted parents were wonderful grandparents to her children, and they were understanding and supportive about Marcy wanting to find her birth mother.

In 2013, Marcy's book club read *Orphan Train* by Christina Baker Kline (HarperCollins, 2013). On a whim, she decided to see how much of the historical fiction was factual, and found a (now-defunct) website to help find birth families. She posted some information about she and her birth mother, and a couple of people tried to help with no results.

Then, a woman named Jessica saw her post and asked for some more of the non-identifying information Marcy had. Almost right away, she found Marcy's birth mother, named Laura. Jessica e-mailed Marcy pictures of Laura's mother's parents (Marcy's grandparents), plus copies of the census with all of the family's names. She even found obituaries for Laura's siblings, along with her married name. With all the good news, Jessica also had some bad: Laura passed away in 2007.

With Jessica's help, Marcy wrote letters to Laura's living sisters and brother. When they didn't respond, Marcy found an address for a second cousin named Samantha and wrote to her as well. A few days later, Marcy got a call from Samantha, who said she didn't understand who Marcy was writing to and who Laura was. She asked Marcy a lot of questions and said she would see if she could find any of the people Marcy told her about. After they hung up, Marcy guessed the caller was pretending not to be her cousin so she could think about what to do.

But an hour later, Marcy got another call—this time from a different Samantha who was excited to hear from Marcy but knew nothing about her. As it turned out, Marcy had sent the letter to the wrong Samantha! But that mistaken Samantha (the one who called Marcy) took the time to find the real one.

Marcy never expected her birth family to be so kind and welcoming, and she felt like she was floating around on the ceiling. She couldn't stop crying and pacing. Marcy wanted to see her cousins that summer and keep in touch, and they shared pictures of Laura and a ton of useful information. Marcy still grieved for a mother she never met, but having her cousins helped fill that hole. After Marcy sent a second letter, an uncle agreed to meet, and he gave her pictures and a hug.

After talking to Samantha, Nancy (one of Laura's sisters) decided to contact Marcy. She told Marcy why Laura left their home state (South Carolina), plus how Laura met her husband, Greg. Nancy even gave Marcy some details about Laura's personality and interests, and shared how close the two sisters were. Marcy still tears up thinking about how much Laura meant to her sister. Sadly, Nancy passed away just a few months later.

Marcy continued reaching out to her birth mother's family, revealing interesting details from each. According to her siblings, Laura hid her pregnancy well—so well, in fact, that they didn't even know she was pregnant, even when she was nine months. But she also learned a piece of information that was confusing: Laura was five month's pregnant when she married, but Marcy's adoption record claims Laura was single at the time. Why would Laura have lied on the adoption record? And who was Marcy's father?

After consulting with her newfound relatives, Marcy put the pieces together. After a big fight with her family, Laura and a friend moved from South Carolina to Michigan, where she met a man named Greg. Before long, Laura and Greg moved to South Carolina, where they got married—and where Laura was five months' pregnant with Marcy. Their stay in the Palmetto State didn't last long, however; the couple moved back to Michigan, where Marcy was born.

Was Greg Marcy's birth father? He seemed to fit the bill, although Marcy couldn't understand why the married couple would have put up their child for adoption—or why Laura listed herself as single on the adoption record. Is it possible Greg was not the birth father, and Laura lied on the record to hide the fact? It was certainly possible, but Marcy need more information to find out for sure. She built a family tree of Laura and Greg's family, and kept looking for answers.

Marcy next turned to DNA. She tested with 23andMe and AncestryDNA (which proved Laura was her birth mother). The latter allowed her to attach her DNA results to Laura and Greg's family tree (see chapter 9 on mirror trees). Since she already knew which of her DNA matches came from Laura's side, Marcy identified some matches as coming from her birth father—and this strongly suggested a relationship to Greg, at least on his paternal side.

Marcy cast a wider net using other testing services. She transferred her autosomal results to Family Tree DNA. There she found a first cousin related to her on Greg's maternal side, which finally gave her the confirmation she needed. From paper records, she knew Greg was the only son on his direct maternal line who survived infancy. As a result, she could confidently determine Greg is her birth father. Marcy finally felt a burden lifted off her shoulders.

Since then, Marcy has found DNA matches to several cousins who are related to Greg, including two of his nephews. She was afraid of what they might say—it's hard to put yourself out there and surprise people with no knowledge of an adoption in the family. Despite her worries, Marcy reached out to the nephews in 2016—her first cousins. One responded right away and sent pictures of Marcy's paternal grandparents, along with other helpful information. Marcy now has pictures and stories about her birth father and her paternal family. Another cousin called Marcy not long after. Everyone had been kind and shared pictures and stories.

As it turned out, Marcy's birth parents were married for over fifty years and are now both deceased. She has so many questions that will never be answered. Thankfully, DNA has erased any doubts about their identities, though Marcy wishes it could answer the questions about what was going on in their lives when they made the decision to place her for adoption. They were good at keeping their secrets, so Marcy will never know if they ever thought about her or had any regrets.

Marcy shared all these findings with her adoptive mother, who was happy for her. Despite all these new revelations, the relationship between them never changed. Marcy and her adoptive mother were close until the moment she passed away.

CASE STUDY

Sue

Sue's story begins with an interesting twist. When some of her paternal grandfather's relatives wanted to know more about their mutual ancestors, Sue began a family tree to research the Goldberg siblings and their parents. Her genealogy research spread in many directions from the nuclear family she originally started with, eventually including spouses, children, and in-laws. But it soon became clear that those who arrived in the Goldberg family had a more pressing issue to explore: a "missing" half-sister on her paternal grandmother's side of the family.

Sue's paternal grandmother, Tillie, died before Sue's parents even met. Sue's father didn't say much about Tillie, other than that she immigrated to New York from Romania. Interested in learning more, Sue began constructing a family tree piece by piece, backed up by her mother who recognized some people and facts that her research uncovered. Sue learned Tillie had two older siblings who also made their way to New York, both of whom arrived at different times in the early 1900s. The two siblings brought spouses and children with them to America.

With this knowledge, Sue discovered many new second cousins and embarked on the emotional journey of connecting with them for the first time. While cautious to approach these relatives and tell them her story, Sue was pleasantly surprised by their positive reception. None of the cousins had really known much about their Romanian grandparents, and were thrilled to learn more.

Two of the second cousins, Stanley and Ann, provided Sue with a great research challenge. Stanley and Ann, both now in their seventies, are the grandchildren of Tillie's brother, Samuel, through Samuel's son (Abe) and daughter-in-law (Helen). After Ann and

Stanley became acquainted with Sue, they divulged they had an unknown half-sister from an earlier marriage of their father Abe, but they had no idea where to find this sister.

Abe was particularly secretive about his first marriage and the child that was born from it. In fact, when Stanley was a child, he approached his father with a picture of a young girl, and Abe rebuked him for questioning him about her. This was, he suspected, the sister.

Over time, Stanley gleaned more information and had some leads to follow. Stanley discovered that Abe's first wife was also named Helen, and this "Helen #1" died shortly after giving birth to the mysterious half-sister. Helen #1's relatives then took over the child's care. And from various hints, Stanley believed she was raised in Delaware by someone who was an engineer. If this baby was still alive, she'd be in her eighties. Still, Stanley's search fell short. He looked for his half-sister for decades, then finally threw in the towel fifteen years ago after exhausting every possible lead.

When Stanley first talked to Sue about it, he sounded pained. He couldn't believe his father would give up his own child—that wasn't the father Stanley had known. However, Sue learned this was common for Jewish families long ago: A widowed man could not keep the baby after the death of his wife. He was expected to work, and was thus unable to care for a baby. Abe may not have had a choice.

Sue was determined to help Stanley find his missing half-sister (who, after all, was also Sue's second cousin). She searched for Helen #1 on Ancestry.com, and found her relatively easily thanks to the site's hinting functionality. Ancestry.com had a marriage record of Abe and Helen—but this Helen had a different maiden name than Stanley and Ann's mother, and the marriage was several years prior to that to their mother's. Then Sue found Helen's siblings and traced them back as best as she could. Sure enough, she found one of Helen's brothers was an engineer and had moved to Delaware. Not everything added up initially, with several inconsistencies in the records she found. But she still felt confident she was moving in the right direction. (And, as we'll see, these inconsistencies eventually made sense.)

Through census data, Sue learned this engineer brother and his wife had a daughter named Arlene who was in the same age range as the half-sister. There seemed to be no record of this daughter's birth, possibly a clue that she was not the biological child of this couple and had arrived in their household later. Additional records supplied a birth date for Arlene, three months prior to Helen #1's death date. Perhaps Helen gave birth to Arlene, then died a few months later (rather than in childbirth, as Stanley surmised)?

Stanley and Ann weren't the only relatives who knew something about Abe's first marriage. Another, one of Abe's nieces, knew Stanley wanted to find his half-sister and provided Sue with a photo of Abe and Helen #1 before their baby was born, beautiful young lovers untouched by tragedy. Abe's children had never seen it—his sister kept it close, but

Abe's niece knew of its existence. Looking at the photo, Sue felt sad—the couple had not spent their lives together as planned, and Abe lost their only child together. Abe's other children simply never met their half-sister, and it was upsetting that this secret was kept so long.

Sue turned her attention to Arlene. In archived records, she discovered that Arlene married twice and was widowed twice. Obituaries for both husbands listed the names of surviving relatives. Arlene and her first husband had three children, but most records for living people aren't available on Ancestry.com. So Sue went on the hunt using Google and Facebook—and, lo and behold, those children were all on Facebook!

Sue saw one of the children had posted a photo of a younger woman and an older woman. One commenter noticed that the younger woman and her "mother" looked great, leading Sue to believe the two were Arlene and her daughter—and that Arlene, the older, was Stanley and Ann's missing half-sister.

Now that she had this information, Sue was troubled by how to approach the family. Did Arlene know her true birth story? Did she know she was raised by her uncle and his wife, rather than by her birth parents? Sue was also unsure how an eighty-four-year-old would receive the information, knowing that other biological relatives were searching for her—would Arlene find the research intrusive, or would she be receptive?

Sue returned to Stanley and Ann to decide what to do next. They laid out the possibilities that Arlene could be happy about the discovery or may turn her back from her half siblings. Stanley and Ann mulled over how to reach out to this potential sister and decided that a letter may be the best way forward. By this time, Sue had also found a phone number and address that appeared to be Arlene's. If they had identified the wrong person and Arlene wasn't their half-sister, Arlene could simply ignore it.

In the letter, the trio took a more conservative approach. They simply said they were researching their family and discovered Arlene's path crossed with theirs, asking if she could help fill in some gaps. They didn't have to wait long—Arlene and one of her daughters, realizing Stanley and Ann were her biological siblings, picked up the phone almost immediately after reading the letter.

Arlene told them she learned she was adopted at age eleven, when a taunting neighborhood boy told her the secret. Her adoptive parents, her uncle (Helen's engineer brother) and his wife, filled in the details and explained the circumstances that led Arlene into their lives. A doctor advised Helen #1 not to get pregnant because of rheumatic heart disease, but she had a baby anyway and died from complications because of it. Arlene went home with Helen's brother in Delaware. Abe visited Arlene several times when he was able to travel from the Bronx—and Arlene even made it to New York to see her father in the Bronx on a few occasions. But the visits stopped around the time Arlene was eight and Abe married Helen #2. Arlene only remembered her "Uncle Abe" via some photos with him from when she was a toddler, plus a locket he had given her. She kept them all these years.

Since Stanley and Arlene lived in the same state, they quickly set up an in-person meeting, hosted by Arlene's daughter. Arlene had never been seen so happy as she was when she reunited with her biological sibling. Though Ann did not live nearby, Arlene and Ann set up a weekly phone call, and Ann is preparing a visit to her at the time of this writing.

For her part, Sue is fortunate to live close to Ann and one of Ann's newly discovered nieces—Sue's second cousin once removed—and all three of them have been getting to know each other. No doubt remained, but the three took DNA tests that confirmed the relationships. Sue had found the missing half-sister.

APPENDIX A

Frequently Asked Questions

You've got DNA questions, and we've got answers. We've covered these topics throughout the book, but I hope you find this quick-reference section helpful.

Does the test hurt?

Not at all! There's no pin-prick, and no blood given. Instead, you swab the inside of your cheek with a cotton swab or spit into a tube. Do note that, in the latter's case, you need to generate a significant amount of spit, so you may want to do it in private.

How do I take the test?

Each testing company provides specific, detailed instructions on how to collect your sample. Saliva-based tests are simple, with samples collected one of two ways: Simply fill a tube with saliva up to a certain line or swab the inside of your cheek with the cotton swab. If you do the former, make sure your sample is filled up to the line without bubbles. To help salivate better, use a tiny bit of sugar or lemon juice.

Which testing service should I use?

The answer to this is mostly subjective and depends on your goals and preferences. In general, test with as many companies as you can afford so you have the best chance of getting in contact with your genetic relatives.

However, if you can only test with one company, make sure you're taking the most appropriate test. Most people will want to start with an autosomal DNA test, which is the test offered by AncestryDNA **<dna.ancestry.com>**, MyHeritage DNA **<www.myheritage. com/dna>**, and 23andMe **<www.23andme.com>**. Family Tree DNA **<www.familytreedna. com>** also offers an autosomal DNA test (called the Family Finder).

Autosomal DNA tests can tell you about both sides of your family, and they are by far the most popular DNA tests. (AncestryDNA, in particular, has more than seven million test-takers in its database.) Each autosomal DNA testing company has its own benefits: AncestryDNA and MyHeritage DNA boast family tree integration, and 23andMe's results include several health-related data points that can clue you into your predisposition for certain traits and diseases.

In addition, Family Tree DNA offers mitochondrial DNA (mtDNA) tests as well as several Y-chromosomal (Y-DNA) tests, though these will have more limited utility for adoptees. mtDNA can teach you about your strictly maternal lines, while Y-DNA tests can teach you about your strictly paternal lines.

Keep in mind that Family Tree DNA and MyHeritage DNA allow you to upload your results from other companies to their databases, allowing you to compare your results to their networks of test-takers. Be sure to take advance of this, as well of third-party resources like GEDmatch **<www.gedmatch.com>** and DNAGedcom **<www.dnagedcom.com>**.

Is my testing data secure?

It's completely valid to be nervous that you've put your DNA results in a database—after all, your DNA is what makes you, you. But take comfort in knowing that security and privacy are of the utmost importance for DNA testing platforms.

Each site has a privacy policy and terms of service that ensure your data will be safe in the company's care. Some service providers will destroy your DNA after you've tested; others will maintain it but associate it—and you—with a bar code or identifier that does not directly identify you by name. Some service providers store your data wholly anonymously in a locked vault so that, if it's broken into, no one can identify which sample is yours. And while the company can use your anonymous DNA in aggregate (e.g., combine it with data from large groups of users), your DNA will never be shared and attached to your name. And if you're still uncomfortable, you can opt out of these studies.

You can also take some proactive steps to protect your DNA data. Some services (such as GEDmatch) allow you to name your DNA results whatever you like, meaning you don't have to use your real name. In fact, you can save your DNA results as "Bubba" or "Ferris Bueller" if you want (though the latter might attract some attention). Be sure to keep any downloaded raw DNA data in a safe place.

My results came in! What next?

This question comes up more often in DNA genealogy groups than you might think. Test-takers often feel overwhelmed, scared, and eager (or reluctant) to know more. Stay calm! You've already committed to learning about your genetic history, and nothing can change the facts about your heritage. Take a deep breath, and remember that this is just the first step on a journey.

Now, log in to the service and view your results. You'll likely see your ethnicity estimate, which (while interesting now) won't be of much use until you're further into your research.

Instead, focus on your DNA matches. Try to make sense of them. Who does the service estimate to be your closest relatives? Do you see any familiar names or places? If you're on AncestryDNA or MyHeritage DNA, examine the family trees of your closest matches, if available. If not, reach out to the user with a simple message of "We appear to be X cousins, and I'd like to know how. Can you help me out?" Consult chapter 9 for more on reaching out to potential relatives. Some users might not be receptive to your messages (and your outreach could always backfire), so be prepared for potential rejection. Gather all the information you can, and you're bound to find a family member.

Continue working down your match list. As you become more familiar with your genetic relatives, start building a mirror tree (see chapter 9) and observing the matches you and your matches have in common.

What's the best test to establish paternity?

To establish paternity, you and your suspected father should take an autosomal DNA test. If the two of you match with 3,500 to 3,600 shared centimorgans (cM), you've found a match! If you're a male, you could also consider taking a Y-DNA test, as you'll share nearly an exact copy of your father's Y-DNA.

How can I tell which matches are maternal or paternal?

If you've had one parent tested (or you've tested a half sibling who you know how you're related to), you're halfway there. Simply look at what matches you and your parent have in common. The matches you and your parent don't have in common are matches through your other birth parent's line, and you can then begin to build out a tree that reflects your newfound relatives. AncestryDNA makes this easy to view, providing a Mother or Father filter on the top of your DNA matches by the search box. On Family Tree DNA, you can also use the In Common With (ICW) tool to isolate matches that are paternal or maternal. (Linking your DNA relative to a your family tree will also become helpful.)

If you know absolutely nothing about either birth parent, this will be more difficult. If you have any fourth cousins or closer, you'll have to work with what you've got. Look at the family trees of your matches and see if you can find common surnames or locations that may match up with where you were born. You can also use this strategy to connect with matches of similar ethnicity, especially if they're close. The same grouping strategy applies if we're using AncestryDNA, with the Shared Match feature that isolates relatives who are likely paternal or maternal. If you're using Family Tree DNA, the ICW tool will give you similar results.

MedBetterDNA, a Google Chrome plug-in available on the Chrome web store, can also help you determine which matches are paternal versus maternal. Once you have it downloaded and installed, MedBetterDNA will allow you to filter your AncestryDNA matches. Simply right click while viewing your AncestryDNA matches, and select which filters you'd like to apply to your match list (such as confidence intervals, starred versus unstarred, and matches that have family trees). You can even create your own filters using hashtags.

Sorting matches as maternal and paternal is not a guarantee, but keep at it!

Which other relatives should I have tested?

Great! You've gotten your DNA tested. Now start convincing other family members to do the same. Adding more DNA results to the database will help you and other test-takers connect with genetic relatives.

In particular, make sure you test older relatives first. Time is working against you, and older relatives are more likely to pass away or otherwise no longer be able to provide samples. In addition, your older matches have more of your ancestors' DNA, meaning you'll be able to make more concrete discoveries about your ancestral DNA.

What is a mirror tree?

If you're an adoptee, you'll often hear about mirror trees. The idea is fully fleshed out in chapter 9, but here's the basic concept: Once you've received your AncestryDNA results, you look for a DNA match's family tree. Then you make a private copy of that tree in your Ancestry.com account and replace the other user as the tree's home person. After some time, AncestryDNA does its magic and identifies potential relatives.

How do I represent both of my families (adopted and birth) on my tree?

Many people, once they've discovered their birth families, want to list both on their tree. There are a few ways to do this. The first is to simply create a tree for your birth family and another for your adoptive family. This will simplify your research, as well as present you with more ancestors to research.

But if you want to combine them somehow, you could also include both on a single family tree by adding "alternate" parents. On Ancestry.com, go to a person's profile page and select Edit Relationships from the Edit dropdown toolbar. Then click Add Alternate Father or Add Alternate Mother to create another slot (image **A**). Enter the person's information, then select the appropriate relationship (e.g., Adopted, Step, Foster) from the dropdown menu. This process can be somewhat cumbersome, however.

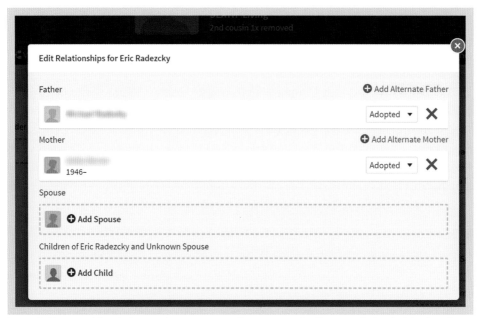

Image A. You can add multiple sets of parents to your Ancestry.com family tree, including both birth and adopted mothers and fathers.

I don't want to make a family tree online. Are there genealogy software programs?

Certainly! Genealogists have turned to offline software programs for decades. My cousin Gary and I recommend RootsMagic **<www.rootsmagic.com>**, which is available for both Mac and PC and boasts error-finding and citation features, plus a variety of export options. Family Tree Maker **<www.mackiev.com/ftm>** is another solid option. And if you change your mind about putting your family tree online, both RootsMagic and Family Tree Maker have syncing capabilities with Ancestry.com and FamilySearch.org **<www.familysearch.org>**.

I can't find any close matches. What can I do? I'm ready to give up.

While DNA has made it much easier for long-lost family members to reconnect, the process isn't immediate or automatic. And it's normal to get frustrated if you haven't found the right match right away, or even after months of searching.

If you've been searching for a long time and have reached your wit's end, simply take a break. It can be heartbreaking to go to the site every day, only to find out that you're still no closer. Set up e-mail notifications for when you receive new matches, or log on less frequently—perhaps every other week. Fortunately, new people are added to the database every day.

Alternatively, you could upload your DNA results to other sites and services. As we discussed earlier, Family Tree DNA and MyHeritage DNA both accept results from other testing companies, opening you up to new potential matches. Services like GEDmatch and DNAGedcom also allow you to compare your results to a different pool of users, plus access additional analysis tools.

Most importantly, keep faith. You're bound to find results, whether it's tomorrow or three months from now. But these tests are becoming more accessible to the general public—not just people who know they've been adopted.

What is shared cM?

Shared cM (short for centimorgan) refers to the amount of DNA you and a match share. Testing services use the amount of shared DNA to determine how closely you and another user are related. Once you know how many cM you share with a match, you can also estimate the relationship yourself by using charts like the ones in chapter 4. Note that multiple kinds of relationships can be indicated by the same amount of cM, so you should look to other factors (such as the length of the largest shared DNA segment) to determine how close you are to a match.

How do I see how many shared cM I have with a match on AncestryDNA?

When viewing a match's page (e.g., to see their tree and or heritage information), click on the little (i) icon next to the confidence level of your match. Under that, a black screen will show up that says "Amount of Shared DNA," which will display your shared DNA in centimorgans as well as how many segments you and the match share. Since AncestryDNA doesn't offer a chromosome browser, this is the most-detailed information representation of your DNA with another user.

How long does it take for test results to come in?

Each service is different. Typically, you'll wait six to twelve weeks, depending on how many other folks have tested recently. However, I've received results back as early as eleven days from the lab receiving the kit (which, in turn, was two or three days after it was mailed). You can monitor your DNA kit's progress on each site, and e-mail notifications will alert you as the kit enters each new phase.

What are the chances of finding my birth parent(s)?

You're in luck: Thanks to the accessibility and price point of genetic testing, you're closer than ever to find your parents. Of course, you can use traditional records to get there faster. But, in the absence of birth or adoption records, DNA testing is a great way to go.

This is sort of a trick question, because there is no answer that fits every adoptee or sperm-donor conceived child. There is no set timeline. Your chance of success depends on which of your biological relatives have tested, and on which of the various services you and your birth family have tested on. Your birth father could already be in Family Tree DNA's system, waiting for you to discover him. But perhaps AncestryDNA is still processing your half-sister's results. Likewise, your biological grandmother (we may not be sure yet if it's maternal or paternal) will be testing on 23andMe in a few months. Close relatives testing is the most important factor in determining how long it will take you to find your birth parents.

But if you have only distant relatives who have tested, you may have to wait longer or do more leg work. While waiting for closer matches to test, you can collaborate with these more-distant relatives—or you can bide your time. If you're from an endogamous people, you're going to have to wait for a closer match.

Thanks to the availability of DNA tests, it's only a matter of time until the right birth relative adds their results to the database.

How do I use the results I have to find a relative?

Use your DNA match list to identify individuals who the site projects are related to you. Then contact those users to ask for access to their family trees and see if they're interested in seeing how the two of you are related. Begin with your closest matches, then move to more distant relatives as you continue to learn more. Be careful when reaching out, though—see chapter 9 for some advice.

More advanced users can also use chromosome browsers to determine what specific pieces of DNA are associated with different branches of your family tree. This can be helpful as you attempt to perform triangulation, where you use information about two known relatives to discover info about a third, unknown relative. See chapters 10 and 11 for more on triangulation.

Why do I have so many AncestryDNA matches, but not any DNA Circles?

AncestryDNA places you in DNA Circles based on your relationships to the people in those Circles, and based on their relationships to a common ancestor and to each other—all built upon the foundation of public family trees. The more people who test, the higher your chances of being placed in a Circle.

You may be encountering one of a few different situations if you're not seeing any DNA Circles yet. Since placement in Circles is dependent on test-taker results with uploaded family trees, your relatives need to have tested at AncestryDNA and uploaded a family tree. If they haven't done either, you won't appear in DNA Circles with them.

AncestryDNA also recommends having a paid subscription to their regular Ancestry.com service. You'll need to make your tree public and expand it as far as you can.

DNA circles can occasionally disappear, typically because a person updates their tree in such a way that it affects the ancestor(s) you share in common. AncestryDNA's algorithm then processes the match and assigns the user to a DNA Circle. Likewise, if you delete the common ancestor, you will get kicked out of the circle.

What cousins are worth researching? Is it too hard to match up when my closest match is in the estimated third cousin (50–60 cM) range?

The answer to this question depends largely on your particular situation. Matches marked as "close relatives" or "first cousins" are almost always worth following up on, but the chances of you being able to pinpoint your relationship to more-distant relatives get murkier from there. You'll generally want to follow up with estimated second cousins, but you'll have mixed results researching estimated third cousins. It's important to look at their trees; these alone can give you a lot of information regarding your shared ancestor.

Can you prove these more-distant matches, some of whom you may share 50 or 60 cM with? Though shared cM values aren't perfect, they can't be ignored—and you can only learn something about your matches if you investigate the relationship. Given the following conditions, you'll have a good chance of discovering the relationship:

- You've already discovered large portions of the family tree
- You don't come from an endogamous community
- You and your match share a large segment of DNA
- The other party is cooperating

Throughout the book, we've talked about other factors that can affect your ability to find a common ancestor with your match: record availability, ethnic heritage, the "uniqueness" of the surnames you're researching, and more.

Why would I share more DNA with someone's child than with either parent?

In general, people share more DNA with a parent than with that parent's child, as autosomal DNA is "lost" from generation to generation. In cases where the opposite is true, you also share DNA with the child's other parent. To put it another way: You share DNA with both of the child's parents. Endogamy could be the culprit, or perhaps the child's parents just happened to share an unusual amount of DNA.

What's a haplogroup, what can it do, and where can I find it?

Haplogroups are groups of individuals who share common DNA markers from a specific geographic location. All members of a haplogroup descend from a population that lived in a particular place thousands of years ago, indicating your "deep ancestry" (i.e., the ancient roots of one branch of your family tree's DNA). You can use haplogroup information to determine if you and another person share DNA along either your maternal (mtDNA) or paternal (Y-DNA) lines. You can learn your haplogroups by taking the appropriate test from Family Tree DNA, or by taking 23andMe's autosomal DNA test.

What is DNA triangulation?

Triangulation is an advanced genetic genealogy technique that involves examining two known DNA kits to determine if they overlap with a third DNA kit on the same chromosome. By using triangulation, you can determine whether a kit comes from your maternal or paternal lines (if known), as well as identify additional relatives.

First, you'll need to find three or more people who match with each other at least 7–10 cM in the same area on the same chromosome. You can do so by viewing your data in GEDmatch.

Let's assume you have four people who appear to triangulate. You can use GEDmatch's Multiple Kit Analysis four times (changing the kit number on the top), or you can use the One-to-One tool six times with the following comparison:

- Comparing kit 1 with kit 2
- Comparing kit 1 with kit 3
- Comparing kit 1 with kit 4
- Comparing kit 2 with kit 3
- Comparing kit 2 with kit 4
- Comparing kit 3 with kit 4

If everyone matches each other on that chromosome in the same place, then you all have the same common ancestor. If not, the matches aren't all related—perhaps some are on your mother's side and the rest on your father's.

In simpler terms, you want to find segments that match in the same location. If kit 1 overlaps with kit 2, kit 2 overlaps with kit 3, and kit 1 overlaps with kit 3 all on the same portion of a chromosome, you have a triangulated match. However, if kit 1 overlaps with kits 2 and 3, but kit 2 doesn't also overlap with kit 3, then the kit 3 match will be from a different ancestor. More on this concept in chapters 10 and 11.

How does the X-match work?

Autosomal DNA is hands-down the most useful kind of genetic genealogy test, but autosomal DNA tests that feature chromosome browsers also include X-chromosomal DNA (X-DNA) data. Like with autosomal DNA matches, you can run comparisons for and view matches of X-DNA relatives.

Similar to Y-DNA, X-DNA is passed down along the sex chromosomes (in this case, the X chromosome). Women have two X chromosomes (XX) while men have one X chromosome and one Y chromosome (XY):

- Women (XX) receive one X chromosome from the father, one from the mother
- Men (XY) receive one X chromosome from the mother, one Y chromosome from the father

You can use X-DNA's inheritance pattern to determine how you're related to X-DNA matches. For example, men always inherit their X-DNA from their mothers, so you can confidently say that a male X-DNA match relates to you along his mother's line. However, this can get trickier when determining how you relate to female X-DNA matches. A woman's X chromosomes come from both parents, making it more difficult which genetic line a segment of X-DNA came from.

The X chromosome recombines less often than the autosomal chromosomes, so a large X-DNA match doesn't necessarily indicate a strong connection. Be sure to look for other evidence (such as a strong autosomal DNA match) to support your theories.

Are there any other tools I should know about?

Yes! Certain Google Chrome plug-ins can easily create Ahnentafel ancestor lists from online family trees, helping you to more easily record yours or other users' family trees. The Ahnentafel format is an ancestor numbering system that assigns each ancestor a distinct number: You receive number 1, your father receives number 2, your mother receives 3, your father's father receives number 4, etc.

If you have trees on Family Tree DNA, check out the plug-in called DNArboretum. Once you download DNArboretum plug-in from the Chrome store, open your family tree on Family Tree DNA and click the DNArboretum icon. This will generate a chart that shows the home person, followed by their parents, grandparents, great-grandparents, etc. Rows are clickable so you can see the direct paths of the ancestors.

A similar tool is Pedigree Thief, also available in the Chrome store. This creates Ahnentafel tables based on Ancestry.com family trees. To use this tool, download the plug-in, then view your Ancestry.com family tree in pedigree view. Navigate to the person who you want to explore. Click on the Pedigree Thief icon and watch the tree animate as it populates the Ahnentafel file for you.

These tools, together with Kitty Munson Cooper's Ahnen2GED <kittymunson.com/dna/Ahnen2GEDcom.php>, are quite helpful for adoptees because they collect trees quickly and easily. You can then upload those trees to a number of available tools (GEDmatch, Wikitree <www.wikitree.com>, or an offline tool of your choosing) and see if ancestors match up.

APPENDIX B

Worksheets

While it can be tempting to jump into research and cousin-matching as soon as you've received your DNA results, you'll benefit more from careful and thorough analysis of your potential matches, web searches, and ancestral lines. This section contains two forms to help you analyze your research and keep your findings in order. The first, the DNA Cousin Match Worksheet, will help you track and compare the names, amounts of shared DNA, and estimated relationships of various potential cousins. The second form (which stretches across two pages) helps you home in on seven specific DNA matches, allowing you to record key information such as shared ancestral places/surnames/ethnic origins, along with estimated relationships and any correspondence.

DNA COUSIN MATCH WORKSHEET

User Name/Kit Number	Percentage Match	Centimorgans (cM)	Relationship	Notes

MATCH RELATIONSHIPS WORKSHEET

	Testing Company and Website	Username of Match	Estimated Relationship	Contact Info (If Known)	Shared Ancestral Places	Match's Ancestors from Shared Places
1						
2						
3						
4						
5						
6						
7						

MATCH RELATIONSHIPS WORKSHEET

	Shared Surnames	Match's Relative(s) with That Surname (and Relationship to User)	Shared Ethnic Origins	Correspondence with User, Including Dates	Notes
1					
2					
3					
4					
5					
6					
7					

INDEX

23andMe, 14, 38, 44, 135
    Ancestry Timeline, 107
    appearance predictors, 102
    chromosome browser, 108
    DNA comparison, 104
    DNA matches, 98–99, 101–102
    downloading raw data, 139
    health and wellness information, 98, 100
    Neanderthal ancestry, 105
    sharing information, 99, 109
    sorting DNA matches, 99
    Strength of Relationship, 101
    triangulating relationships, 101–104
    viewing results, 99, 101–104
23andMe reports
    ancestry composition, 106–108
    haplogroups, 108–109, 112
    your DNA family, 109, 112
529andYou, 110–111

Adopted.com, 26
Adoptee Rights Law Center, 25
adoption laws regarding birth records, 23, 25
adoption registries, 25–26
    age as limiting factor, 25
    list by state, 30–32
Adoption Reunion Registry, 26
adoption, closed, 25
alleles, 37
American Adoption Congress, 25
ancestry errors, avoiding, 127
Ancestry.com
    calculating estimated relationship, 129

commenting on family trees, 130
    family trees, 63, 73
    Map and Locations, 75
    member directory, 130
    messaging system, 73
    mirror trees, 124–127, 129–130
    Pedigree and Surnames, 73–74
    Shared Matches, 74, 129
AncestryByDNA, 64
AncestryDNA, 14, 38, 44, 50, 135
    DNA circles, 74–75
    DNA matches, 68–75
    downloading raw data, 138
    ethnicity estimates, 63–68
    matching criteria, 68, 71
    reference populations, 65
    testing process, 62–63
    viewing data, 72–75
ancientOrigins, 87–88
Ashkenazi Jews, 34, 39, 66, 152, 154
autosomal DNA
    Family Tree DNA Family Finder test, 79–89
    inheritance patterns, 37–39
    statistics chart, 46
    testing, 34, 37–39, 44–48, 79–89
    See also 23andMe; AncestryDNA; Family Tree DNA; MyHeritage DNA
autosomal DNA segment analyzer (ADSA), 172–176
autosomes, 36

BeenVerified, 28
Bettinger, Blaine T., 47
Big Y test, 60–61, 76

birth certificates, 22–23, 25

    access to by state, 23

    from different countries, 25

birth families, reasons for seeking, 13–14

birth surnames, 61

blogs, 27

centimorgans (cM)

    and endogamous communities, 50

    shared, 16, 45, 72–73, 79, 104, 140, 146–148

    Shared cM Project, 48–49

Chromosome Browsers, 72

    23andMe, 108, 170

    Family Tree DNA 81, 83–85, 172, 179

    GEDmatch, 147–148, 150

    MyHeritage DNA, 120–122, 169, 172

Chromosome Painting, 154

chromosomes, 36

Clinical Laboratory Improvement Amendments (CLIA) certification, 38

cloning, 40

cM. See centimorgans (cM)

collaboration, 76, 131–134

cousinhood, 16–17

    genetic vs. genealogical, 35

deep ancestry, 87, 89

deoxyribonucleic acid. See DNA

Discoverly, 29

DNA

    basics of, 35–37

    See also autosomal DNA; DNA testing; mitochondrial DNA (mtDNA); X-chromosome DNA (X-DNA); Y-chromosome DNA (Y-DNA)

DNA Circles, 74–75

DNA data

    downloading, 138–139

    running diagnostics, 146

DNA Detectives, 27–28, 131

    Autosomal Statistics Chart, 46

DNA inheritance, 38–39

DNA matches

    23andMe, 98–99, 101–102

    AncestryDNA, 68–75

    Family Tree DNA, 80

    MyHeritage DNA, 118–120

    troubleshooting, 68–70

DNA Painter, 170, 172

DNA Quest project, 120

DNA testing

    autosomal, 34, 37–39, 44–48, 79–89 (see also 23andMe; AncestryDNA)

    available tests, 14

    Big Y, 60–61, 76

    changes in design layouts, 62

    confronting skeptics, 19

    cotton swab, 39–40, 76

    expectations of, 18–20, 71

    full vs. half-identical regions, 47

    lab analysis, 40–41

    limitations of, 41–42

    mitochondrial DNA, 44, 53–54, 56–58, 67, 89–92

    overseas, 76

    predicting relationships, 71–73, 79–80

    process of, 39–41

    reasons for, 15, 18

    retaking the test, 40

    saliva sample, 39–40, 62–63

    security issues, 40, 117

    short tandem repeat (STR), 60

    using after traditional research techniques, 21

    when results fail, 34

    X-DNA, 44, 50, 52, 67

Y-DNA, 44, 58–61, 76, 92–97

See also 23andMe; AncestryDNA; Family Tree
DNA; MyHeritage DNA

DNA triangulation, 85

DNAGedcom, 172–177

Dodecad, 152

Double Match Triangulator (DMT), 178–180

endogamous communities, 34, 48, 68, 71, 141

and mirror trees, 125

and shared cM, 50

estimated relationships, calculating, 129

Ethiohelix, 152

ethnicity estimates

23andMe, 106–108

AncestryDNA, 63–68

cautious approach to, 37

Family Tree DNA, 86–87

MyHeritage DNA, 115–118

ethnicity

African, 152

Asian, 152

Eurasian, 152

European, 152

Jewish, 66

Middle Eastern, 152

multiethnic, 67

Native American, 67

South Asian, 152

unexpected results, 63–64, 67

See also Ashkenazi Jews

Eurogenes, 152

eye color, 41–42

Facebook groups, 27, 131. *See also* social media

Family Finder (FF) tool, 79

Family Tree DNA, 14, 38, 44, 56, 61, 76, 135

accessing results, 77

Advanced Matches, 97

Ancestral Surnames, 80

ancientOrigins, 87–88

autosomal DNA test, 76

Big Y test, 94

certificates, 91, 95

Chromosome Browser, 81, 83–85

DNA matches, 80

downloading raw data, 138–139

ethnicity estimates, 86–87

Family Finder test (autosomal DNA), 76, 79–89

home page, 78–79

In Common With (ICW) tool, 85–86

Linked Relationships, 80

linking matches to family members, 82

Matches Maps, 94–96

matching criteria, 80–81

the Matrix, 87–89

mitochondrial DNA test, 76, 89–92

mtDNAFullSequence test, 89

mtDNAPlus test, 89

myOrigins, 86–87

other tools, 97

testing process, 76

uploading data from, 175

uploading family trees, 82

uploading other test results, 81

viewing results, 77–79

Y-DNA tests, 76, 92–96

family trees

creating, 63

communicating on, 130

searching both sides, 168

uploading, 82

FindMe.org, 26

full identical regions (FIRs), 47

G's Adoption Registry, 26

GEDCOMs, uploading to GEDmatch, 160

GEDmatch, 15, 133, 135

    3D chromosome browser, 147–148, 150

    Admixture (heritage) tool, 152–154

    Admixture with Population Search, 149

    Chromosome Painting, 154–155

    comparing three kits, 150

    Generations matrix, 150

    getting started, 136

    multiple kit analysis, 150

    people who match one or both of 2 kits, 145–146

    personal tools, 150

    Phasing, 144–145

    raising the threshold, 142

    uploading GEDCOMs to, 160

GEDmatch comparison tools

    One-to-Many matches, 136–137, 140–141

    One-to-One compare, 141–143

    X One-to-One, 143–144

GEDmatch, Tier 1 tools

    Lazarus, 160

    matching segment search, 157

    My Evil Twin Phasing, 166

    one-to-many matches, 156

    relationship tree projection, 158–159

    triangulation, 161–163

    Triangulation Groups, 163–165

GedrosiaDNA, 152

genealogy

    common mistakes, 127

    paper trail, 169

genes, 37

genetic conditions, 100

genetic distance, 56

genetic genealogy, 56

genetic heritage, 19

Genetic Information Nondiscrimination Act, 40

genetic relatives

    contacting, 19, 24, 75, 79, 81–82, 109, 131–134

    contacting via social media, 27–28

    creating mirror trees, 124–130

    precautions before contacting, 73

GenetiConcept, 135

Genome Mate Pro (GMP), 110, 177–178

Google+, 27

habits, 41

half-identical regions (HIRs), 47

haplogroup information

    in 23andMe test results, 98, 108–109, 112

    in Family Tree DNA results, 89–97

    map, 90

    migration, 55–56, 68, 91, 112

    from mtDNA testing, 44, 53–58, 68, 89–92, 108–109, 140

    for Native American tribes, 67

    phylogenetic tree, 55

    projects, 76

    from Y-DNA testing, 44, 60–61, 93, 95–96

HarappaWorld, 152

health and wellness information, 15, 98, 100

    inclination toward health conditions, 41

homozygosity, 154

human genome, 41, 57, 91–92

In Common With (ICW) tool, 85–86

inheritance patterns, autosomal DNA, 34, 37–39

Instagram, 27, 29

International Society of Genetic Genealogy (ISOGG), 50, 92

International Soundex Reunion Registry, 26

Judaism, as religion/ethnicity, 66

JWorks, 176–177

KWorks, 176–177

Lazarus, 160

LinkedIn, 27

location information, 85

Map and Locations, 75

Matches Maps, 94–96

Matching Segment Search, 157

maternal line, 54, 89

Matrix, 87–89

MDLP Project, 152

medical information, 25

migration
    groups, 55, 68
    maps, 68, 91
    patterns, 88

mirror trees, 73
    on Ancestry.com, 124–127, 129–131
    and endogamy, 125

mitochondria, 35

mitochondrial DNA (mtDNA)
    compared to X-DNA, 52
    Family Tree DNA test, 89–92
    and genetic distance, 56
    and haplogroup migration, 44, 53–58, 68, 89–92, 108–109, 140
    inheritance descendants chart, 53

    interpreting results, 57
    and the maternal line, 54
    phylogenetic tree of, 55
    testing, 44, 53–54, 56–58, 67, 89–92

Mitochondrial Eve, 54, 108

most recent common ancestor (MRCA), 168

My Evil Twin Phasing, 166

MyHeritage DNA, 14, 38, 44, 113, 135
    chromosome browser, 120–122
    dashboard, 114
    DNA matches, 118–120
    downloading raw data, 139
    ethnicity estimates, 115–118
    match list, 114
    overview, 115
    shared ancestral surnames, 120
    uploading results from other companies, 115
    viewing results, 114–115, 118–120

MyHeritage Family Tree, 118

myOrigins, 81, 86–87

Native American DNA, 67

Neanderthal ancestry, 105

non-paternity events, 13, 18

nucleus, 35

One-to-Many Matches, 157

online search tools, 28–29

Oracle, 154

paternal line, 58–59, 92

paternity determination, 18, 61

Pedigree and Surnames, 73–74

Phasing, 144–145, 166

phylogenetics, 55–56

physical characteristics, 41–42

Pinterest, 27

Pipl, 28

Predict Eye Color tool, 41–42

privacy concerns, 40, 117

psychological information, 25

puntDNAL, 152

QuickBase Adoption Database, 26

recombination, 38, 52

Reconstructed Sapiens Reference Sequence (RSRS), 57, 91

reference groups/panels, 64–65

Relationship Tree Projection, 158–159

Reunion Registry, 26

search angels, 131–132

search engines, 28

Search Squad, 27–28, 131

search strategies,

    adoption registries, 25–26

    birth certificates, 22–23

    before DNA testing, 21

    online, 28–29

    social media, 27–28

Shared cM Project, 48–49

Shared Matches, 74, 129

shared surnames, 74, 81–82, 120

short tandem repeat (STR) tests, 60

single nucleotide polymorphisms (SNPs), 41, 85, 94, 156

social media, 14, 26–28

    reaching out to relatives through, 27–28

surname projects, 76

triangulation, 101–104, 161–163

    basics, 167–170

Triangulation Groups, 163–166

triangulation tools

    Autosomal DNA Segment Analyzer (ADSA), 172–176

    DNA Painter, 170, 172

    DNAGedcom, 172–177

    Double Match Triangulator (DMT), 178–180

    Genome Mate Pro, 177–178

    JWorks, 176–177

    KWorks, 176–177

    The Triangulator, 181–183

Twitter, 27, 29

WeGene, 135

X-chromosomal DNA (X-DNA)

    compared to mtDNA, 52

    female inheritance, 51

    male inheritance, 51

    testing, 44, 50, 52, 67

Y-Chromosomal Adam, 60, 108

Y-chromosomal DNA (Y-DNA), 44, 58–61, 76

    Big Y test, 60–61, 76

    Family Tree DNA tests, 92–96

    haplogroups, 44, 60–61, 93, 95–96

    inheritance descendants chart, 59

    standard Y-STR values, 96

    testing, 44, 58–61, 76, 92–97

Y-DNA match map, 95

PHOTO CREDITS

COVER

Cover image: Caiaimage/Sam Edwards Collection:OJO+, Getty Images

PART 1

Opener: Caiaimage/Sam Edwards, Collection: Caiaimage, Getty Images

CHAPTER 1

How to Calculate Cousinhood sidebar: Courtesy Family Tree Magazine <www.familytreemagazine. com>

CHAPTER 2

Image A: Courtesy Jason Greenberg. Used with permission

Image B: Based on image courtesy the Adoptee Rights Law Center <adopteerightslaw.com>. Used with permission

Image C: Courtesy Facebook <www.facebook.com>. Used with permission

CHAPTER 3

Image A: Kateryna Kon, with edits from Tamar Weinberg. Purchased from Shutterstock

Image B: Courtesy GEDmatch <www.gedmatch. com>. Used with permission

PART 2

Opener: Credit:Andrew Brookes, Collection: Cultura, Getty Images

CHAPTER 4

Image A: Based on an image courtesy Family Tree DNA <www.familytreedna.com>

Image B: Courtesy the DNA Detectives <thednadetectives.com>. Used with permission

Images C and D: Courtesy Blaine T. Bettinger <thegeneticgenealogist.com>. Used with permission

Images E and F: Based on images courtesy Blaine T. Bettinger <thegeneticgenealogist.com>. Used with permission

Image G: Based on an image courtesy Charles F. Kerchner, Jr. "Genetics & Genealogy - An Introduction" <kerchner.com/books/introg&g. htm>

Image H: Based on images courtesy Family Tree DNA

Image I: Courtesy Wikipedia <en.wikipedia.org/wiki/ Human_mitochondrial_DNA_haplogroup>

Image J: Courtesy Wikimedia Commons. Created by User:Maulucioni

Images K and L: Courtesy Family Tree DNA. Used with permission

Image M: Based on an image courtesy Charles F. Kerchner, Jr. "Genetics & Genealogy - An Introduction"

Image N: Courtesy Family Tree DNA. Used with permission

Image O: Courtesy Wikimedia Commons. Created by User:Y-dna data file

CHAPTER 5

All images courtesy Ancestry.com <www.ancestry. com>. Used with permission

CHAPTER 6

All images courtesy Family Tree DNA <www. familytreedna.com>. Used with permission

CHAPTER 7

Images A–H: Courtesy 23andMe <www.23andme. com>. Used with permission

529andYou sidebar: Courtesy 529andYou <www.529andyou.com> by Roger Woods. Used with permission

CHAPTER 8

All images courtesy MyHeritage <www.myheritage. com>. Used with permission

PART 3

Opener: Weekend Images Inc., Collection:E+, Getty Images

CHAPTER 9

All images courtesy Ancestry.com. Used with permission

CHAPTER 10

Images A–S: Courtesy GEDmatch <www.gedmatch.com>. Used with permission

Download Your Raw Data sidebar: Images courtesy (clockwise, from top-right to center-left) Family Tree DNA, 23andMe, MyHeritage, and Ancestry.com

CHAPTER 11

Images A–B: Courtesy DNA Painter <dnapainter.com> and Jonny Perl. Used with permission

Images C–E: Courtesy DNAGedcom <dnagedcom.com>. Used with permission

Image F: Courtesy Genome Mate Pro < getgmp.com>. Used with permission

Image G: Courtesy Louis Kessler. Used with permission

Images H–I: Courtesy Göran Runfeldt. Used with permission

Images J–L: Courtesy RootsFinder

PART 4

Opener: bernie_photo, Collection: iStock, Getty Images Plus

CASE STUDY: DONNA

Image A: Courtesy Mark Strauss via GEDmatch. Used with permission

CASE STUDY: KALANI

Images A–B: Courtesy Kalani Mondoy. Used with permission

Q&A

Image A: Courtesy Ancestry.com. Used with permission

Acknowledgments

There are a lot of people I can thank for helping me write this book. First, there are the inspirations behind my pursuit: my cousins Gary Palgon and Jason Greenberg, the family genealogists. It was my cousin Franceksa who got me into adoption research, and online Facebook groups like DNA Detectives and Search Angels as well as DNA Help—Your Jewish Journey and Tracing the Tribe—Jewish Genealogy on Facebook that inspired me to want to learn more about my own history and to help others. It was cousins like Steve Turner and friends like Alan K'necht who really compelled me to go the extra mile. It was passionate advocates like Jennifer Mendelsohn who worked tirelessly to make DNA easily understood by everyone, especially those in the Jewish space which has challenges of its own.

Then, there are the experts in their craft who have truly inspired me to learn more and created platforms (or participated within them) to give me the ammunition and the knowledge to write a book on adoption and DNA less than two years after I really got into the subject matter. It all took many, many hours and effort to catch up, but the resources they provide are aplenty: Blaine T. Bettinger, who has been writing about DNA for quite a long time and whose own book compelled me to go deeper into more niche topics (like adoption and, eventually, endogamy within my own community); Kitty Munson Cooper, who also builds up a passionate following with her blog and her resources, especially her chromosome mapper which gives beautiful visuals of the people who have contributed to your genetic makeup—and who can carry on a conversation for hours on the topic with her fellow genealogists, even if there's a three-year-old crying in the background; CeCe Moore, who has been an inspiration in the community; Roberta Estes, Barbara Shoff, and Robin Bauer, whose writings helped me better understand quite a lot when I was starting out; Louis Kessler for his awesome Double Match Triangulator product; Jonny Perl for creating the amazing DNA Painter tool; the DNAGedcom and Genome Mate Pro teams for providing countless hours to make DNA genealogy easier and more accessible; Dawn Kosmakos, who was always lending an ear when I was challenged by questions from the experts; and Deborah Castillo, who goes above and beyond to simplify DNA to anyone in complete paragraph form. I'd also like to acknowledge my family: Max Heffler, who seems as determined as I am to figure out how we're connected (and with the data in front of us, we should know!); Max Preston, who is also digging through some incredible archives to make pretty significant strides in our genealogical research; the late Mildred Edelman z"l, who shared stories of a great-grandfather I never had a chance to meet; Ruth Tubero; Molly Harris; Bruce Gerber; Helen Dow; Gloria Gruber; Sy Siegal; Ruth Sabo; Berel-Anne Evans; Bruce Decter; Doug Kaplan; Meryl Futersak; Irwin Bernstein; George Jochnowitz; Ken Barkin; Keith Pitzele; Jeff Nobel; Avie Rock; Marcia Markowitz; Francine and Neil Strauss; Marcia Pepper; Tom and Allen Levine; Steve Fisher; Mikki Bergman; and Gail Hochman; my many other cousins who have helped me take my passion of our family genealogy to the next level by sharing their own stories, even if just a little, and collaborating; and the Search Angels who teach me so much every day and have motivated me to create a manuscript that I hope serves the adoption community well.

I'd like to thank Andrew Koch at F+W for giving me a chance to write this—and on that note, since I didn't know better when I wrote my first book (in 2009), I owe a thanks to Colleen Wheeler for her editorial insights when I was at it the first time. (And in case you were wondering, no, that book wasn't about DNA or adoption.)

Finally, I want to thank the numerous individuals who have granted me permission to use images in this book and who have given me the help I needed to write this thing, including those who have shared their own stories. If not for Kristie Wells, formerly of Ancestry, I doubt I'd be here. Today, Jessica Murray at Ancestry continues to hold the torch to keep that program going. Thank you both. Also, a big thanks to Bennett Greenspan, CEO at Gene By Gene/Family Tree DNA, who has not hesitated to get on the phone with me for both brainstorming and ideas, and who never hesitated to answer an e-mail—often immediately.

Without any of you, this book would be an idea in my imagination.

About the Author

Tamar Weinberg is a professional hustler, genealogy enthusiast, and writer, having authored the bestselling book *The New Community Rules: Marketing on the Social Web*. She's appeared on *CBS News* and *The Agenda with Steve Paikin*, and has been quoted in *Forbes*, *USA Today*, *BusinessWeek*, and many other publications. She is an active participant in online communities relating to genealogy and DNA, and after thousands of hours of study and research, has helped solve many cases of unknown parentage. She resides in New York with her husband and their four children. Read more about Tamar on **<tamar.com>**.

Dedication

This book is dedicated to my ancestors and descendants, my incredibly supportive husband Brian, and my four beautiful children, David, Sarah, Alana, and Danny. To Mom and Dad, Grandma and Grandpa, and the multitude of uncles, aunts, and cousins from near and far who gave me their tests to analyze further which gave me many different perspectives on DNA and where I came from. And finally, to Jackie Morris, whose relationship may never be determined, but who I know has some relationship to me somewhere. As far as I'm concerned, you're my littlest sister even if we don't share full identical regions.

4 FREE Family Tree Templates

- decorative family tree posters
- five-generation ancestor chart
- family group sheet
- relationship chart
- type and save, or print and fill out

Download at <www.familytreemagazine.com/freeforms>

MORE GREAT GENEALOGY RESOURCES

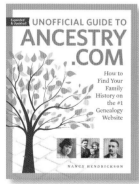

Unofficial Guide to Ancestry.com

By Nancy Hendrickson

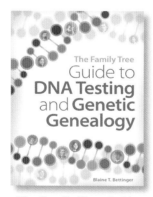

The Family Tree Guide to DNA Testing and Genetic Genealogy

By Blaine T. Bettinger

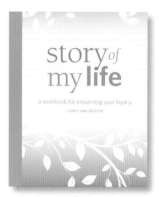

Story of My Life: A Workbook for Preserving Your Legacy

By Sunny Morton

Available from your favorite booksellers and **<familytreemagazine.com/store>**, or by calling (855) 278-0408.

Join our community! <facebook.com/familytreemagazine>